Cyclodextrins in Chromatography

RSC Chromatography Monographs

Series Editor: Roger M. Smith, *University of Technology, Loughborough, UK*

Advisory Panel: J.C. Berridge, *Sandwich, UK*; G.B. Cox, *Illkirch, France*; I.S. Lurie, *Virginia, USA*; P.J. Schoenmaker, *Amsterdam, The Netherlands*; C.F. Simpson, *London, UK*; G.G. Wallace, *Wollongong, Australia*.

This series is designed for the individual practising chromatographer, providing guidance and advice on a wide range of chromatographic techniques with the emphasis on important practical aspects of the subject.

Supercritical Fluid Chromatography
edited by Roger M. Smith, *University of Technology, Loughborough, UK*

Packed Column SFC
by T.A. Berger, *Berger Instruments, Newark, Delaware, USA*

Chromatographic Integration Methods, Second Edition
by Norman Dyson, *Dyson Instruments Ltd, UK*

Separation of Fullerenes by Liquid Chromatography
edited by K. Jinno, *Toyohashi University of Technology, Japan*

HPLC: A Practical Guide
by Toshihiko Hanai, *Health Research Foundation, Kyoto, Japan*

Applications of Solid Phase Microextraction
edited by Janusz Pawliszyn, *University of Waterloo, Ontario, Canada*

Capillary Electrochromatography
edited by Keith D. Bartle, *University of Leeds, UK*
and Peter Myers, *X-tec Consulting Ltd, UK*

How to obtain future titles on publication

A standing order plan is available for this series. A standing order will bring delivery of each new volume upon publication. For further information please contact:
Sales and Customer Care, Royal Society of Chemistry, Thomas Graham House, Science Park, Milton Road, Cambridge CB4 0WF
Telephone: + 44(0) 1223 420066

RSC
CHROMATOGRAPHY
MONOGRAPHS

Cyclodextrins in Chromatography

Tibor Cserháti and Esther Forgács

Chemical Research Centre, Hungarian Academy of Sciences, Budapest, Hungary

advancing the chemical sciences

Chemistry Library

ISBN 0-85404-540-6

A catalogue record for this book is available from the British Library

Published by The Royal Society of Chemistry,
Thomas Graham House, Science Park, Milton Road,
Cambridge CB4 0WF, UK

For further information see our web site at www.rsc.org

Typeset by Keytec Typesetting Ltd, Bridport, Dorset, UK
Printed by Athenaeum Press Ltd, Gateshead, Tyne and Wear, UK

Preface

Chromatographic separation techniques are principally based on the partition of analytes between stationary and mobile phases of different polarity. According to the characteristics of the stationary and mobile phases and the forces moving the mobile phase, chromatographic methods can be divided into gas solid (GSC), gas–liquid (GLC), thin-layer (TLC) and high-performance liquid (HPLC), super-critical fluid (SFC) chromatography and electrically driven separation methods, such as capillary electrophoresis (CE), capillary gel electrophoresis (GCE), micellar electrokinetic capillary chromatography (MEKC), capillary isoelectic focusing (CIEF), and capillary isotachophoresis (CITP).

Cyclic oligosaccharides (cyclodextrins, CDs) can form inclusion complexes with a considerable number of organic and inorganic compounds. The formation of inclusion complexes can markedly modify the physicochemical parameters of the guest molecule (adsorption capacity, polarity, hydrophobicity, *etc.*). As the same physicochemical parameters govern the retention of analytes in a chromato-graphic system, the formation of inclusion complexes influences the retention behaviour of solutes. As the complex formation may modify the behaviour of individual analytes differently, the use of cyclodextrins in chromatography may result in modified retention behaviour and consequently better separation.

As the strength of the complex formation between CDs and positional and optical isomers is generally different, the beneficial effect of CDs on the separation of these isomer classes is of paramount importance.

Contents

CHAPTER 1

Chemistry and Physicochemistry of Cyclodextrins

The main products of photosynthesis are two glucose polymers: cellulose and starch. These two carbohydrate polymers represent a considerable part of the renewable energy source of the Earth. While cellulose is the resistant and structure-forming component in the cells, starch is the mobilisable, convertible energy depot. Starch consists of two glucose polymers: amylose and amylopectin. Dextrins, the products of partial hydrolysis of starch, can give cyclic dextrins under the action of glucosyltransferase enzyme. Cyclodextrins (CDs) are cyclic, non-reducing oligosaccharides consisting of D-glucopyranose units bonded through α-1,4 linkages.

The three major cyclodextrins are the following; the smallest is the α-CD (Schardinger's α-dextrin, cyclomaltohexose, cyclohexaglucan, cyclohexaamylose or C6A) with six glucose units, followed by β-CD (Schardinger's β-dextrin, cyclomaltoheptose, cycloheptaglucan, cycloheptaamylose or C7A) with seven glucose units and γ-CD (Schardinger's γ-dextrin, cyclomaltooctose, cyclooctoglucan, cyclooctaamylose or C8A) with eight glucose units. Unsubstituted native CDs are crystalline, non-hygroscopic, homogeneous substances, which are torus-like macrocycles built up from glucopyranose (glucose) units (Figure 1.1). In CDs, the sugars adopt a $^4C_1$ chair conformation and orient themselves so that the molecule forms a toroidal truncated cone structure. The cavity is lined by the hydrogen atoms and the glycosidic oxygen bridges. The non-bonding electron pairs of the glycosidic oxygen bridges are directed towards the inside of the cavity to produce a high electron density and lend it some Lewis base character.

The C-2 hydroxyl group of one glycopyranose unit can form a hydrogen bond with the C-3 hydroxyl group of the neighbouring glucopyranose unit. In the β-CD molecule a complete secondary belt is formed by these hydrogen bonds, so that β-CD is a rather rigid structure. This arrangement of hydrogen bonds can explain the observation that β-CD has the lowest solubility of all native CDs. The lower solubility of β-CD in water relative to α-and γ-CDs may also be due to the marked structure of water arising from water–β-CD interactions.

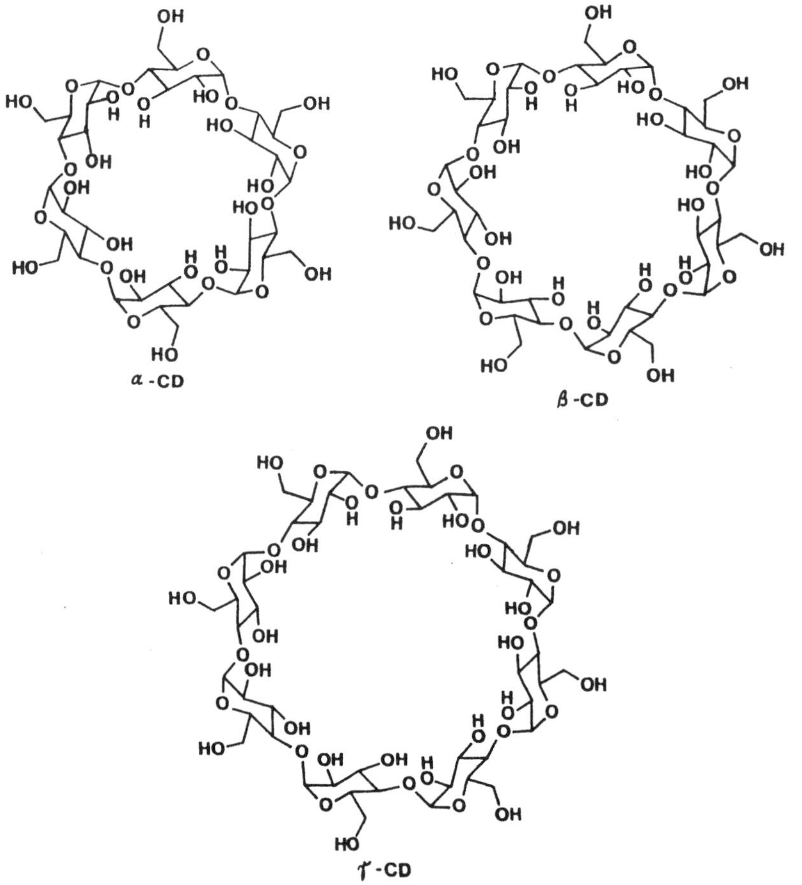

Figure 1.1 *Schematic structures of native α-, β- and γ-CDs*
(Reprinted with permission from ref. 1)

The hydrogen belt is incomplete in the α-CD molecule because one of the glycopyranose units is in a distorted position; therefore, instead of the six possible hydrogen bonds, only four can be formed. The γ-CD molecule has a more flexible structure and consequently is the most soluble of the three native CDs. The equilibrium constants for hydrogen–deuterium exchange in the secondary hydroxyl groups of α-, β- and γ-CDs also indicate that the strongest hydrogen bond system is formed in the β-CD molecule. The most important physical and chemical characteristics of native CDs can be compared (Table 1.1).

Because of their torus-like geometry, relatively hydrophobic surface of the internal cavity and the hydrophilic character of external hydroxyl groups (Figure 1.2), CD molecules easily form inclusion complexes with a wide variety of organic and inorganic molecules. This complex-forming capacity is the reason for their widespread application in chemistry and in separation science (Table 1.2).[1]

Table 1.1 *Physical and chemical characteristics of native cyclodextrins*

	α-CD	β-CD	γ-CD
Glucose units	6	7	8
Internal diameter (nm)	0.47–0.53	0.60–0.65	0.75–0.83
Depth of cavity (nm)	0.79	0.79	0.79
pK_a value	12.3	12.2	12.1
Water solubility*	145	18.5	232
MW (DA)	972	1135	1297
Approximate volume of cavity**	174	262	427
Crystal water (%,w/w)	10.2	13.2–4.5	8.13–17.7
Hydrophobic interaction	CD-cavity	CD-cavity	CD-cavity
Hydrogen bond	glucose-OH	glucose-OH	Glucose-OH

* (mg mml⁻¹, 25 °C); ** (10⁶ pm³).
  (Reprinted with permission from ref. 12)

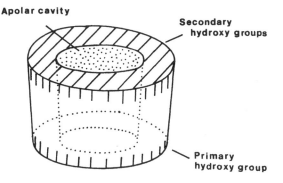

Figure 1.2 *Functional scheme of CD torus*
        (Reprinted with permission from ref. 1)

Table 1.2 *The use of cyclodextrins in analytical separation methods*

Chromatographic method	Mode of use of CD in the system
GSC	Deposit on inert support or immobilized
GLC	(chemically bonded) selective component of a liquid stationary phase
HPLC	Chemically bonded stationary phase or
TLC	Additive to the mobile phase
SFC	Chemically bonded stationary phase
CZE	Additive to the background electrolyte
MEKC	Charged CD derivatives as transport agents/additive to micellar electrolyte systems
ITP	Additive to the leading electrolyte

(Reprinted with permission from ref. 1)

CDs can be created with larger cavity diameters; however, these products are interesting only from a synthetic point of view, because they can contain too many water molecules, so that the driving force of complex formation is negligible.

1 Selector Properties

As has been frequently demonstrated cyclodextrins are capable of forming inclusion complexes with compounds having a size compatible with the dimensions of the cavity. Inclusion complexes are entities comprising two or more molecules; the 'host' includes a 'guest' molecule, totally or in part, by only physical forces, that is, without covalent bonding. CDs are typical host molecules and may include a great variety of molecules having the size of one or two benzene rings, or even larger compounds, which have a side chain of comparable size, to form crystalline inclusion complexes. Complexation occurs when there is a steric compatibility between the CD cavity and the guest molecule and the affinity of the guest molecule for the CD cavity is higher than for the other components present (*i.e.* solvent). It has been established that, besides steric compatibility, hydrophobic interactions,[2] van der Waals interactive forces[3] and hydrogen bonding[4] independently or in combination play a considerable role in the determination of the strength of inclusion complexes. The stoichiometry of inclusion compounds is usually 1:1; however, complexes can be made of two or more guests (especially with the large γ-CD cavity) or with several CD molecules by the inclusion of different parts of a large guest molecule.

Many physicochemical methods have been used for the study of the formation of inclusion complexes. Thus calorimetry,[5] spectrophotometry[6] and various liquid chromatographic methods, such as reversed-phase thin-layer and high-performance liquid chromatography,[7] have been successfully used for the assessment of the various aspects of host–guest interaction.

Inclusion complex formation is stereoselective; thus it affords a possibility of resolving enantiomers. A reasonable, simple three-point interaction model[8] has been adopted for explanation of the enantiomer separations of CDs in liquid chromatography. This model is based on the assumption that one part of the host molecule, preferably an aromatic group, must interact with the cavity of the CD and the other parts of the molecule can interact with the primary and the secondary OH groups of the CD rim. As the steric matching and consequently the energy of interaction between the two enantiomers and the three points of attachment can be different, it may result in the separation of the corresponding enantiomer pair yielding enantiodifferentiation. Molecular mechanistic investigations[9] have also been used to explain the behaviour of CDs and CD derivatives as chiral phases or chiral stationary phase additives in GC. The calculations indicated a marked opening of the rim of the primary OH groups in the cyclodextrin's cavity and of the secondary OH groups being tilted inwards to the cavity (presumably to enhance the network of secondary hydrogen bonding).

It has also been deduced that the remarkable flexibility of the CD ring can be responsible for the adaptation of molecules to a wide range of shapes.

CDs have been effectively employed for chiral separations and for the improvement of a large number of separation processes in many chromatographic techniques, either as mobile-phase additives[10-12] (dissolved in the mobile phase, as modifiers) or as stationary phases or stationary-phase additives.[13-15] CDs offer numerous advantages. They are thermostable to a reasonable temperature which is important in GC; furthermore, they are stable over very wide pH range (2-12) and do not absorb radiation in the region normally associated with UV detection (200-350 nm) facilitating their application in liquid chromatographic techniques.

2 Solubility of Cyclodextrins

The water solubility of native cyclodextrins shows anomalous behaviour (see Table 1.1). The solubility of β-CD is only 1.85 g/100 mL at ambient temperature, whereas the solubilities of α-CD and γ-CD are significantly higher, 14.5 g/100 mL and 23.2 g/100 mL, respectively. The temperature dependences of the solubility of the three native CDs in water can be described by the following equations:[16]

$$\alpha\text{-CD}; \quad c = (112.71 \pm 0.45)e^{-(3530\pm31)[(1/T)-(1/298.1)]} \tag{1.1}$$

$$\beta\text{-CD}; \quad c = (18.3236 \pm 0.099)e^{-(14137\pm31)[(1/T)-(1/298.1)]} \tag{1.2}$$

$$\gamma\text{-CD}; \quad c = (219.40 \pm 9.8)e^{-(3187\pm320)[(1/T)-(1/298.1)]} \tag{1.3}$$

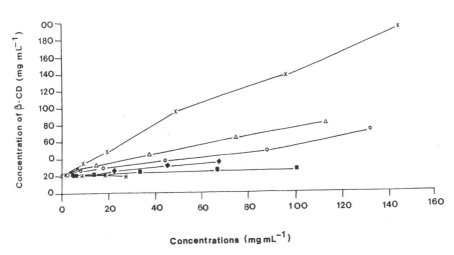

Figure 1.3 *Solubility of β-CD in presence of citric acid, DL-tartaric acid, ascorbic acid, lactic acid, malic acid and hydrochloric acid*
(Reprinted with permission from ref. 16)

where c is the concentration of CD ($mg\,mL^{-1}$) (the values of the constants are followed by their standard errors) and T is the temperature in Kelvin. According to Equations 1.1–3 the solubility of the CDs depends strongly on the temperature: at 50 °C the solubility of all three CDs is about three times higher than at 20 °C.

In the presence of organic molecules, the solubility of native CDs generally decreases owing to complex formation (Figure 1.3).

3 Dielectric Properties

Dielectric constants of CD cavities have been established from the change in the fluorescence properties of occluded pyrene-3-carboxaldehyde. The estimated dielectric constants of CD cavities have been calculated as 48 and 55 for β- and γ-CDs, respectively.[17] These values can be explained by the assumption that the larger γ-CD cavity can possibly include more water molecules, thus providing a larger effective dielectric constant. The dielectric constants of native CD cavities on incorporating the toluidinyl group of 6-p-toluidinylnaphthalene-2-sulfonate at pH 5.3 and 25 °C have also been determined and were found to be 47.5, 52.0 and 70.0 for α-CD, β-CD and γ-CD, respectively.[1] It has been further established that the dielectric constant of the CD cavity depends not only on the inner diameter of the cavity but also on the size of the included guest molecule.

4 Thermal Properties

CDs and CD derivatives have been extensively used as stationary phases or stationary-phase additives in GC. As GC analyses are generally carried out at elevated temperatures, an exact knowledge of the thermal stability of CDs and CD derivatives under the GC conditions is of paramount importance. A considerable number of thermoanalytical methods have been employed for the study of thermostability of CDs and their corresponding inclusion complexes.[18]

The thermal degradations of native cyclodextrins and substituted β-cyclodextrins were characterised by hot-stage microscopy (HSM), differential scanning calorimetry (DSC), thermogravimetry (TG) and X-ray powder diffractometry (XRD). The measurements indicated that cyclodextrins have no well-defined melting point, but from about 197 °C they begin to decompose. It was further found that in an inert atmosphere each of them decomposed in a single major step (250–400 °C) leaving a residue which is thermally quite stable at higher temperatures. Furthermore, it has been frequently shown that the thermal stability of CD derivatives depends on the type, location and number of any substituents. In the case of amino- and phosphate-substituted cyclodextrins, the first degradation step is the same as for native CDs, then volatile by-products are formed by oxidation below 600 °C, the residue being a ceramic-like substance stable to 800 °C. The charring process involves opening the CD rings followed by a chemical decomposition similar to that of cellulose with loss of the glucosidic structure and hydroxyl groups and a build up of carbonyl groups and aromatic structures. Carbon dioxide, water, levoglucosan and furans are the major volatiles evolved from cyclodextrin degradation in the same manner as from cellulose. Products derived from substituted CDs show that they do not simply behave as leaving groups but rather take part in the charring process.[19,20]

Although chromatographic techniques mainly employ derivatised CD and not native CDs the number of studies dealing in detail with the thermal properties of CD derivatives is surprisingly low. This phenomenon is probably because CDs are used for practical purposes in chromatography (to improve separation of enantiomers and other hard to separate analytes) and the exact determination of the thermal properties of CDs has been considered as of secondary importance.

As previously mentioned CDs are not hygroscopic, but form various stable hydrates. Figure 1.4 illustrates the water sorption and desorption isotherms of native cyclodextrins.[21]

CDs have no reducing end groups. They give positive reactions in test characteristic for non-reducing carbohydrates. Therefore, they give a colour reaction with anthrone making possible their quantitative determination by spectrophotometric methods.

Rate constants for the hydrolysis of linear and cyclic dextrins determined at various temperatures and acid concentrations have been compiled (Table 1.3). The results indicate that by decreasing the concentration of the acid at a fixed temperature, or by lowering the temperature at a fixed acid concentration, differences in the hydrolysis rates of cyclic and linear dextrins can be modified.

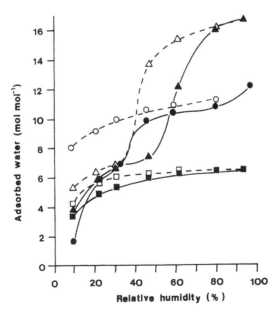

Figure 1.4 *Water vapour sorption isotherms for α- (□, ■), β- (○, ●) and γ-CD (△, ▲)*
showing adsorption (-) and desorption (−)
(Reprinted with permission ref. 25)

Table 1.3 *Rate constant for the acid hydrolysis of β-CD*

Temperature (°C)	HCl concentration (M)	$10^3 \, k_{lin} \, (min^{-1})$	$10^3 \, k_{cycl} \, (min^{-1})$
100	1.15	82.0	48.0
	0.115	5.1	2.7
	0.0115	0.27	0.13
80	1.15	8.2	3.7
	0.115	0.430	0.19
60	1.15	0.75	0.27
40	2.30	0.140	0.060
	1.15	0.0320	0.010
26	5.00	0.37	0.20

Constants for the final phase are characteristic of linear dextrins, more precisely of maltose (k_{lin}), k_{cycl} the value characteristic of the ring opening (hydrolysis) of CD.
(Reprinted with permission from ref. 1)

5 Chemically Modified Cyclodextrins (Cyclodextrin Derivatives)

In CDs every glucopyranose unit has three free OH groups, two of which (on C-2 and C-3) are secondary and one (C-6) primary.[22] As each of these free hydroxyl groups can be modified, by substituting the hydrogen atom or the hydroxyl group by a wide variety of substituents, the majority of simple synthetic reactions results in a considerable number of positional isomers. Hydroxyl protons can be exchanged by deuterium at the oxygen–hydrogen bonds or carbon–hydrogen bonds in deuterated CD derivatives. These molecules represent the smallest group of CD derivatives and have found application only in NMR studies of CDs. Some inorganic esters of CDs, such as nitrates, sulfates, phosphates *etc.* have also been synthesised; however, the organic (acetyl, benzoyl, propionyl, methyl and carbamoyl) esters have been more frequently used for practical industrial and analytical purposes.

Ether derivatives are the most important CD derivatives from a practical point of view. These compounds can be prepared either by direct reaction with an alkylating agent or *via* an intermediate, such as sulfonate esters or deoxy-halogeno derivatives. This group equally contains neutral, anionic and cationic derivatives as well as silyl ethers.[23]

Deoxy cyclodextrins can be classified according to the mode of preparation into two groups: intermediaries which can be further derivatised, and end products containing thio-, amino-, substituted amino- or azido-substituents.

Branched cyclodextrins (or second-generation cyclodextrins) can also be obtained by chemical synthesis but in most cases they are prepared by enzymatic reactions. Branched CDs can be divided into two categories. Homogeneous branched cyclodextrins have only glucose or malto-oligosaccharide side chains bound to the native CDs. Heterogeneous branched CDs have one or more

Table 1.4 *Most common substituents used in derivatised CD*

Methylated β-CD
Acetylated α-, β-, γ-CD
Carboxymethylated β-CD
Hydroxypropylated β-CD
Naphthylethyl carbamate β-CD
3,5-Dimethylphenyl carbamate β-CD
p-Toluoyl β-CD
α-Methylbenzylamine modified β-CD
Naphthylethylamine modified β-CD
Pyridylethylene diamine β-CD
Nitropyridylethylene diamine β-CD
Phenyl carbamate β-CD
Cationic β-CD

galactose or mannose residues bonded either to each other or directly to the parent cyclodextrin rings. The solubilities of branched cyclodextrins in water, even in aqueous 80% ethanol or in aqueous 50% solutions of methanol, formaldehyde and ethylene glycol, are extremely high in comparison with their parent CDs.[24,25]

It is well known that derivatisation changes the physical and chemical properties of the CDs, modifying their solubility, complex-forming capacity, thermal properties and chemical stability. Numerous CD derivatives have found application in chromatography (Table 1.4).[26] The aim may be the improvement of the solubility of the native CDs making possible an increase in their concentration in a chromatographic mobile phases, the enhancement of the stability of the host–guest inclusion complexes by modifying the reactivity and/or mobility of the guest molecule. Furthermore, derivatisation can be employed for the formation of insoluble and/or immobilised CD-containing structures, which can be used as stationary phases for liquid and gas chromatography.

6 Cyclodextrin Polymers

These compounds consist of two or more cyclodextrin rings covalently bonded to each other using a spacer molecule. According to the synthetic pathway they can be classified into linear polymers, which are generally water soluble, and cross-linked polymers, which are water insoluble. The water soluble CD polymers have a higher solubility than their parent cyclodextrins. Many examples have been reported of the solubilising effect of cyclodextrin polymers for poorly soluble drugs.

As with native and derivatised CDs, these polymers have also found applications in chromatography. Water soluble polymers have been used as mobile phase additives in liquid chromatographic techniques and in electrically driven separations.[27,28]

References

1. J. Szejti and T. Osa (Eds), *Comprehensive Supramolecular Chemistry*, Vol. 3, 1996, Elsevier Science, New York, USA.
2. B. V. Müller and E. Albers, *J. Pharm. Sci.*, 1991, **80**, 599–604.
3. M. Suzuki, M. Kajtár, J. Szejtli, M. Vikman, E. Fenyvesi and L. Szente, *Carbohydr. Res.*, 1991, **214**, 25–33.
4. J. H. Park, M. D. Jang and M. J. Sain, *J. Chromatogr.*, 1992, **595**, 45–52.
5. G. Castronuovo, V. Elia, D. Fessas, A. Giordano and F. Velleca, *Carbohydr. Res.*, 1995, **272**, 31–40.
6. Y. L. Loukas, E. A. Vyza and A. P. Valiraki, *Analyst*, 1995, **120**, 333–338.
7. T. Cserháti and E. Forgács, *Anal. Biochem.*, 1997, **246**, 206–209.
8. C. F. Dalgliesh, *J. Am. Chem. Soc.*, 1952, **74**, 3940–3943.
9. K. B. Lipkowitz, *J. Org. Chem.*, 1991, **56**, 6357–6367.
10. A. Italia, M. Schiavi and P. Ventura, *J. Chromatogr.*, 1990, **503**, 266–271.
11. M. Gazdag, G. Szepesi and L. Huszár, *J. Chromatogr.*, 1986, **351**, 128–135.
12. J. Zukowski, D. Sybilska and J. Bojarski, *J. Chromatogr.*, 1986, **364**, 225–232.
13. J. H. Maguire, *J. Chromatogr.*, 1987, **387**, 453–458.
14. S. M. Han, Y. I. Han and D. W. Armstrong, *J. Chromatogr.*, 1988, **441**, 376–381.
15. T. Cserháti and E. Forgács, *J. Chromatogr. A*, 1996, **728**, 67–73.
16. M. J. Jozwiakowski and K. A. Connors, *Carbohydr. Res.*, 1985, **143**, 51–55.
17. G. S. Cox, N. J. Turro, N. C. Yang and M. J. Chen, *J. Am. Chem. Soc.*, 1984, **106**, 422–424.
18. C. Brauer, M. P. Merlin and T. Guerandel, *J. Inc. Phenom. Macrocyc. Chem.*, 2000, **37**, 75–82.
19. F. Trotta, M. Zanetti and G. Camino, *Polym. Degrad. Stab.*, 2000, **69**, 373–379.
20. G. P. Benetti, F. Giordano, V. Massarotti, A. Gazzaniga and P. Mura in *Minutes of the 5th International Symposium on Cyclodextrins*, Paris, ed. D. Duchene, Editions de Sante, Paris, 1987, pp. 2–45.
21. T. Steiner, S. A. Maison and W. Saenger, *J. Am. Chem. Soc.*, 1991, **113**, 5676–5678.
22. D. W. Armstrong, *J. Liq. Chromatogr.*, 1980, **3**, 895–899.
23. P. Fugedi, *Carbohydr. Res.*, 1989, **192**, 366–369.
24. Y. Mizobuchi, M. Tanaka and T. Shono, *J. Chromatogr.*, 1980, **194**, 153–157.
25. M. Yamamoto, A. Yoshida, F. Hirayama and K. Uekama, *Int. J. Pharm.*, 1989, **49**, 163–167.
26. D. Duchene, *New Trends in Cyclodextrins and Derivatives*, Editions de Sante, Paris, 1992.
27. F. Bressole, M. Audran, T. N. Pham and J. J. Vallon, *J. Chromatogr. B*, 1996, **687**, 303–336.
28. T. Cserháti, A. Dobrovolszky, E. Fenyvesi and J. Szejtli, *JHRC&CC*, 1983, **442**, 35–37.

Use of Cyclodextrins in Gas–Liquid Chromatography

Various cyclodextrin derivatives have been used in the GC analysis of a wide range of organic compounds. The overwhelming majority of applications are for the separation of optical isomers because of the growing demands of both industrial and legislation organisations. The growing demand for pure chiral compounds and for efficient control methods have made necessary the development of GC techniques suitable for the chiral separation of a wide variety of compounds, such as pesticides, other environmental pollutants, pharmaceuticals *etc*. An extensive search for new synthetic routes for CD derivatives has been motivated by the relatively low thermal stability of the unmodified CDs and by the recognition that each CD derivative shows different separation characteristics but that only a limited number of chiral separations can be performed on any single CD stationary phase.

1 Determination of the Strength of the Host–Guest Interaction

The number of theoretical studies dealing with the measurement of the enthalpy and entropy of the interaction between host and guest molecules, and the elucidation of the influence of the various binding forces on the strength of interaction is surprisingly low. A theoretical study used native, 3-methylated and 1,3-dimethylated 5-methyl-5-(C_nH_{2n+1}) hydantoin derivatives as model compounds for the study of the influence of molecular parameters on the retention and selectivity of the enantiomer separations.[1] Fused silica capillary columns coated with heptakis-(2,3,6-tri-*O*-methyl)-β-CD (perMe-β-CD), 20% perMe-β-CD and 80% OV-11 (35% phenyl and 65% methyl), and 20% perMe-β-CD and 80% OV-1701 (7% phenyl, 7% cyanopropyl, 85% methyl and 1% vinyl) were used. The enthalpy and entropy of the analyte stationary-phase interaction were calculated from the dependence of the retention on the temperature. The

Table 2.1 *Thermodynamic parameters deduced from GC experiments for native, 3-methylated and 1,3-dimethylated 5-methyl-5-(C_nH_{2n+1})-hydantoin derivatives. The $H°_1$, $S°_1$, and $G°_1$ refer to interactions between the first eluted solute and the perMe-β-CD; ($H°$), ($S°$) and ($G°$) values are related to the enantiomeric discrimination*

n	$H°_1$ kJ mol^{-1}	$S°_1$ J mol^{-1} K^{-1}	$G°_1$* kJ mol^{-1}	$(H°)$ kJ mol^{-1}	$(S°)$ J mol^{-1} K^{-1}	$(G°)$* J mol^{-1}
a) Native: 5-methyl-5-(C_nH_{2n+1})-hydantoin derivatives						
2	−91.2	−120.2	−39.1	0.0	0.0	0
3	−93.9	−124.2	−40.1	0.0	0.0	0
4	−96.9	−128.0	−41.5	−0.7	−1.3	−96
5	−100.8	−133.5	−43.1	−1.0	−1.8	−169
6	−105.3	−139.6	−44.8	−1.1	−2.1	−185
7	−110.6	−147.6	−46.7	−1.0	−2.0	−175
8	−116.0	−155.8	−48.6	−1.1	−2.1	−176
9	−120.3	−161.6	−50.3	−1.2	−2.4	−189
b) 3-Methyl-5-methyl-5-(C_nH_{2n+1})-hydantoin derivatives						
2	−76.2	−105.7	−30.5	−1.7	−3.5	−164
3	−79.3	−110.5	−31.4	−0.8	−1.6	−69
4	−81.4	−112.2	−32.8	−1.0	−2.2	−32
5	−84.8	−116.4	−34.4	−0.9	−1.8	−94
6	−90.6	−125.4	−36.3	−1.8	−3.8	−192
7	−96.6	−135.0	−38.1	−2.9	−6.1	−227
8	−104.2	−148.3	−39.9	−2.2	−4.7	−189
c) 1,3-Dimethyl-5-methyl-5-(C_nH_{2n+1})-hydantoin derivatives						
2	−66.6	−94.7	−25.6	−1.9	−3.9	−204
3	−63.0	−85.8	−25.8	−3.2	−6.7	−284
4	−66.3	−90.6	−27.1	−1.7	−3.4	−203
5	−70.0	−94.9	−28.9	−1.3	−2.6	−156
6	−76.8	−106.1	−30.9	−1.4	−2.9	−161

At 160 °C

(Reprinted with permission from ref. 1)

thermodynamic parameters are compiled in Table 2.1. Native hydantoins were separated better on mixed columns whereas methylated derivatives were more effectively separated on the pure perMe-β-CD column. The chiral selectivity was explained by the influence of hydrophobic interactions, formation of hydrogen bonds and van der Waals forces between the analytes and the stationary phases.

A computational study was carried out to elucidate the binding site and the interactive forces between chiral alkanes, alcohols, and acetates to permethylated β-CD.[2] It was found that chiral recognition also depends on the short-range van der Waals forces and that the enantiodifferentiating forces are markedly smaller (1–2 orders of magnitude) than the binding forces.

2 Separation of Optical and Positional Isomers with Cyclodextrins

Separation of Positional and Optical Isomers of Pesticides

Because of the large number of pesticides used in modern agricultural practice, many GC methods have been developed and applied for their chiral separation and quantitation. Particular attention has been paid to halogenated compounds and the separation properties of different stationary phases for a range of matrices. The enantiomeric separation of organochlorine compounds in biological matrices has been reviewed.[3]

The performance of heptakis-(6-*O*-TBDMS-2,3-di-*O*-methyl)-β-CD 50% dissolved in OV1701, FS-Hydrodex-β-3P and CP-Chirasil-Dex CB β-CD bound to dimethylpolysiloxane was compared for the enantiomeric separation of polychlorinated pesticides.[4] A typical chromatogram on the last column shows the good separation characteristics of the CD-coated stationary phase (Figure 2.1), and it was stated that this good separation can also be exploited in the enantioselective analysis of this class of pesticides in fatty samples.

Chlordane components and metabolites have also been analysed with high-resolution GC using a β-CD derivative as chiral selector and electron-capture negative ion (ECNI) mass spectrometric detection.[5] Enantioselective separations were performed on a fused capillary column coated with OV1701 containing 35% heptakis-(2,3-dimethyl-6-*t*-butyldimethylsilyl)-β-CD (TBDM) (Figure 2.2). The

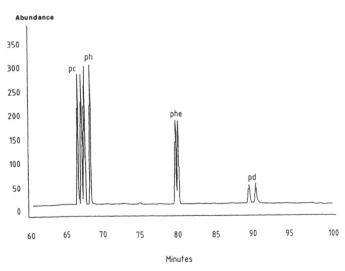

Figure 2.1 *Separation of enantiomers of photo-heptachlor (ph), photo-heptachlorepoxide (phe), photo-dieldrin (pd) and photo-chlordane (pc) give description of Column 3 and of operating conditions*
(Reprinted with permission from ref. 4)

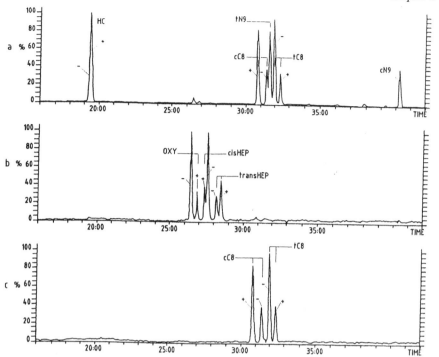

Figure 2.2 *HRGC ECNI chromatograms of an enantio-enriched reference mixture using an OV1701-TBDM column showing the elution of (a) HC = heptachlor; octa- and nonachlordanes (m/z 300), (b) OXY = oxychlordane; cis- and trans-HEP = cis- and trans-heptachlorepoxide (m/z 282), (c) cis- and trans-chlordane (m/z 410). Note the resolution of the chiral compounds (except heptachlor) into enantiomers. Enantiomer elution sequence as indicated (For conditions, see text)* (Reprinted with permission from ref. 5)

separation capacity was compared with columns coated with heptakis-(2,3, 6-O-trimethyl)-β-CD (PMCD) and randomly silylated heptekis-(O-t-butyldimethylsilyl)-β-CD (BSCD). The chromatograms clearly showed that the best enantiomeric resolution can be achieved on the TBDM column, the other two CD derivatives showed inferior separation properties.

Separation of Toxaphenes

GC methods have also been extensively applied for the separation of the components of technical toxaphene (polychlorinated bornanes). Enantiomeric separation of persistent compounds in toxaphene were separated on a *t*-butyldimethylsilylated β-CD (β-BSCD) stationary phase.[6] Enantiomer separation of bornane congeners from a Soxhlet extraction of a hake liver (*Merluccius merluccius*) and two whitebaked dolphin blubber samples (*Lagenorynchus albirostris*) was achieved by heart-cut multidimensional gas chromatography using

ECD detection.[7] This method improved the efficiency of separation by transferring the relevant parts of chromatogram obtained on an Ultra 2 (5%-phenyl–95%-methylsilicone) to an OV1701-TBDM (9:1, w/w) for further separation (Figure 2.3). It was stated that the technique allowed the enantiomer separation and interference-free quantification of bornane congeners even in complicated biological matrices.

The thermal decomposition of toxaphene congeners on various stationary phases was studied in detail.[8] It was established that the peak areas depended on the character of the stationary phase, heating rate and chemical structure of the toxaphene congener and that TBDMS-CD was the preferred column. The enantiomeric forms of the contact insecticide bromocyclene were also separated on a 50% TBDMS-CD–50% OV-1701 capillary column, and bromocyclen was detected by ECD.[9] It was found that the enantiomeric ratio of bromocyclen in river water is always near to one indicating that microbiological decomposition of bromocyclen in water is negligible. The origin of the bromocyclen could be traced back to the waste water of sewage plants where the concentration of bromocyclen varied between 3300 and 50 000 pg L^{-1}. Enantiomeric ratios in fish muscles considerably deviate from zero. This finding was tentatively explained by the assumption that the metabolism and/or accumulation of bromocyclen enantiomers may be different in fish muscles.

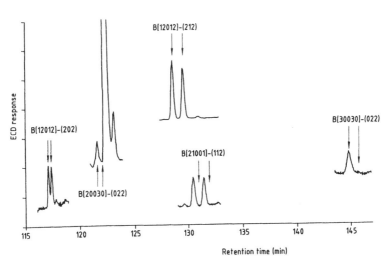

Figure 2.3 *Portions of the heart-cut chromatograms of B[20030]-(022), att. 4; B[12012]-(212), att. 1; B[21001]-(112), att. 2 and B[30030]-(022), att. 4 from the dolphin blubber sample no. 1 and B[12012]-(202), att. 8 from the dolphin blubber sample no. 2. B[20030]-(022) = 2-exo-5,5,9,9,10,10-heptachlorobornane; B[12012]-(212) = 2-endo,3-exo,5-endo,6-exo,8,8,9,10,10-nonachlorobornane; B[21001]-(112) = 2-exo,3-endo,6-endo,8,9,10,10-heptachlorobornane; B[30030]-(022) = 2,2,5,5,9,9,10,10-octachlorobornane; and B[12012]-(202) = 2-endo,3-exo,5-endo,6-exo,8,8,10,10-octachlorobornane* (Reprinted with permission from ref. 7)

Elution orders and enantiomer resolution of organochlorine compounds obtained on four CD stationary phases [heptakis-(2,3,6-tri-*O*-methyl)-β-CD mixed with dimethylpolysiloxane (1:9 w/w) (β-PMCD), heptakis-(2,3,6-tri-*O*-methyl)-β-CD mixed with dimethylpolysiloxane (1:9 w/w) (β-PMCD), β-TBDM-OV1701 35:65, w/w) and 10% heptakis-2,3,5-tri-*O*-*t*-butylmethylsilylated β-CD in 85% dimethyl-, 15% diphenyl-polysiloxane (β-BSCD$_g$)] were compared for selected organochlorine pesticides in marine biota (Table 2.2).[10,11] The data indicate that the separation capacity of the columns show considerable differences even between batches, making necessary the testing of each column before analysis. It was assumed that the inhomogeneity of TBDMS preparations accounts for the differences observed. The method has been successfully used for the determination of the enantiomeric ratios of organochlorine compounds in liver of polar fox (*Alopex lagopus*), and blubbers of harbour seal (*Phoca vitulina*), grey seal (*Halichoerus gripus*) and Caspian seal (*Phoca caspica*)[10] and for herring oil (Figure 2.4).[11]

Toxaphene congeners have been separated on a chiral stationary phase based on dimethyl-*t*-butylsilylated heptakis(2,3,6-tri-*O*-methyl)-β-CD and the method has been employed for the measurement of congeners in blubber of seal. Enantioselective accumulation was not observed.[12] The composition of commercial toxaphenes has also been analysed by enantioselective GC using various CD derivatives for chiral separation.[13] The method has been successfully employed for the measurement of toxaphene components in tissues of aquatic vertebral species (herring, salmon, seal and penguin). It was established that toxaphene components are very resistant to biological decomposition.[14] Toxaphene congeners have also been determined in fish and monkey adipose tissue and in human milk using terbutyldimethylsilylated-β-CD as chiral selector and the enantioselectivity of metabolism has been observed.[15] The influence of the type and composition of

Table 2.2 *Enantiomer resolution and elution order of cis- and trans chlordane on the tested TBDMS-CD capillaries. See text for column description*

	Enantiomer resolution					Elution order
Capillary	γ-HCH	HEP	cis-Chlordane	trans-Chlordane	Tox 50	cis (cs)- and trans (tr)-chlordane
a	1.5	1.8	2.1	1.7	1.4	cs(−),tr(+),cs(+),tr(−)
b	1.7	2.7	0.95	n.s.	1.0	cs(−),cs(+),tr(−)
c	1.5	2.0	1.2	n.s.	0.8	cs(−),cs(+),tr(−)
d	1.7	2.0	1.5	4.6	2.0	tr(+),cs(+),cs(−),tr(−)
e	4.0	1.6	3.3	1.8	n.s.	cs(+),cs(−),tr(−),tr(+)
f	0.7	1.1	0.7	2.1	0.9	tr(+),tr(−),cs(+),cs(−)
g	n.s.	2.3	0.7	2.3	0.8	tr(+),tr(−),cs(−),cs(+)

n.s.: not separated; γ-HCH: γ-hexachlorocyclohexane; HEP: hepta chlor-*exo*-epoxide; Tox 50: 2-*exo*,3-*endo*,5-*exo*,6-*endo*-8,8,9,10,10-nonachlorobornane
(Reprinted with permission from ref. 10)

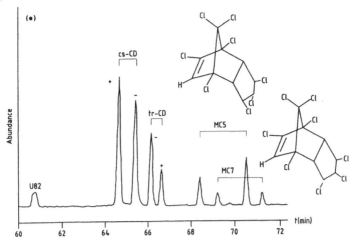

Figure 2.4 *Separation of chlordane isomers U82, cis-, trans-chlordane, MC5 and MC7 in a herring oil sample on the capillary TBDMS-CD. The GC-NICI-MS mass chromatograms of m/z 408 are shown (fused-silica capillaries 15–25 m × 0.25 mm ID). The temperature program was optimised for the best separation of MC5 and MC7. 82: Chlordane of unknown structure, cs-CD, tr-CD: cis-, trans-chlordane*
(Reprinted with permission from ref. 10)

the CD-based stationary phases on the efficacy of chiral separation has been studied in detail and considerable differences in the separation capacity have been esablished.[16,17] The components of chlordane, its metabolites and photoconversion products have also been separated by chiral GC using native CDs, permethylated, perethylated and *t*-butyldimethylsilylated derivatives. Enatiomeric ratios in fish, seal and penguin proved the existence of enantioselective decomposition.[18]

Separation of HCH

Enantioselective preparative GC has been employed for the chiral separation of xenobiotics[19] on columns coated with heptakis-(2,6-di-*O*-methyl-3-*O*-pentyl)-β-CD in polysiloxane OV 1701 (20%). Enantiomers of α-HCH and pentachloro-cyclohexane have also been separated by capillary GC using permethylated β-CD in DB 1701,[20] and the method has been applied to the study of the decomposition of α-HCH. A similar GC technique has been used for the separation of perdeuterated and unlabelled α-HCH on a CP-Chirasil-DEx column.[21] Permethylated CD capillary columns served for the measurement of the enantiomeric ratio of α-HCH in a small Arctic lake and its watershed and the enantioselectivity of the decomposition was observed.[22] The enantioselectivity of the decomposition of α-HCH has been proven also in the Bering-Chukchi seas and the Arctic ocean using capillary columns coated with 20% permethylated

β-CD in methylphenylpolysiloxane and 30% *t*-butyldimethylsilylated-β-CD in methylphenylcyanopropylpolysiloxane.[23] The metabolism of HCHs in Arctic waters has been observed in another experiment using capillary column coated with 20% γ-CD.[24] Permethylated β-CD stationary phase (20%) was employed for the study of the enantioselective degradation of HCH isomers in sewage sludge under anaerobic conditions.[25] The enantioselectivity of the decomposition of HCH isomers has been established also in marine biota using 50% octakis-(3-*O*-butyryl-2,6-di-*O*-pentyl)-γ-CD and 50% heptakis-(2,3,6-tri-*O*-pentyl)-β-CD in OV 5701.[26] The enantioselective decomposition of α-HCH has also been followed in the brain and other issues of neonatal Northern fur seals (*Callorhinus ursinus*), and marked differences in the degradation rate of isomers were observed. Chiral separations have been carried out on capillary columns coated with permethylated-β-CD and heptakis-(3-*O*-acetyl-2,6-di-*O*-pentyl)-β-CD.[27] A combined GC–HPLC method has been developed and applied for the determination of the absolute configuration of (+)-α-1,2,3,4,5,6-hexachlorocyclohexane in the blubber and brain of harbour seal (*Phoca vitulina*) using various CD derivatives as chiral selectors.[28] Organochlorine pesticides have also been measured in harbour and grey seals from western Iceland by GC and the isomers of α-HCH have been separated on CD stationary phases.[29] The accumulation ratio of α-HCH isomers has been determined in marine mammals from the northern hemisphere using β-CD based chiral selectors. It was established that the isomeric ratio strongly depended on the species.[30] The enantioselective transformation and accumulation of cyclodiene pesticides in marine and terrestrial biota was assessed by GC using heptakis-(2-*O*-methyl-3,6-di-*O*-pentyl)-β-CD as chiral selector.[31] Similar method has been applied for the determination of the enantiomeric ratios of oxychlordane and α-HCH in harbour (*Phoca vitulina*) and grey seals (*Halichoerus grypus*). Isomers were separated on a capillary column coated with 35% heptakis-(6-*O*-*t*-butyl-2,3-di-*O*-methyl)-β-CD in OV 1701. The enantioselective accumulation has been proven again.[32] The enantiomeric ratios of α-HCH also showed high diversity in the blubber of small cretaceans. Chiral separations were performed on a β-CD-based stationary phase.[33] The decomposition of α-HCH and cyclodiene in roe deer (*Capreolus capreolus*) liver has been studied by chiral GC and the enantioselectivity of the decomposition has been proven. The stationary phase consisted of 50% octakis-(3-*O*-butyryl-2,6-di-*O*-pentyl)-γ-CD and 50% OV 1701.[34]

Tandem columns have also been employed for the simultaneous analysis of chlordanes and *o,p'*-DDT in environmental samples. Achiral and chiral columns were coated with 90% biscyanopropyl, 10% phenylcyanopropylpolysiloxane and dimethyl-*t*-butyl-silylated β-CD dissolved in 85% methyl 15% phenylpolysiloxane, respectively. The method was proposed for the analysis of marine biota and ambient air.[35] Enantiomers of bromocyclen have been separated from fish tissue on a capillary column coated with 50% heptakis-(6-*O*-*t*-butyl-dimethylsilyl-2,3-di-*O*-methyl)-β-CD and 50% OV 1701. It was found that the muscle tissue of rainbow trout, orfe, bream and pike contains a considerable amount of bromocyclen enantiomers.[36]

The microbial decomposition of chiral endosulfan metabolites was followed by enantioselective GC carried out on heptakis-(2,3,6-tri-*O*-pentyl)-β-CD and hepta-kis-(3-*O*-acetyl-2,6-di-*O*-pentyl)-β-CD. It was observed that the decomposition pathway differed under aerobic and anaerobic conditions.[37]

Separation of Positional and Optical Isomers of Other Environmental Pollutants

Modified CDs have also been employed for the separation of other environmental pollutants. Thus, the separation capacity of OV 1701 and OV 1701 modified with 10% heptakis-(2,3,6-tri-*O*-methyl)-β-CD was compared using polychlorinated biphenyls (PCBs) as model compounds.[38] It was established that the retention order of analytes changes markedly in the presence of modified β-CD (Figure 2.5). Atropisomeric PCBs were also separated on a fused-silica capillary column coated with 25% β-BSCD in 15% diphenyl, 85% dimethylpolysiloxane.[39] The good separation power of the method makes it suitable for the atropisomer separation of PCBs even in biological matrices. Not only PCBs but also enantiomers of chlorinated bis-propyl ethers were separated on various chiral stationary phases containing modified CDs[40] coated with the following chiral stationary phase: 50% heptakis-(6-*O*-*t*-butyldimethylsilyl-2,3-di-*O*-methyl)-β-CD in OV 1701; 50% octakis-(3-*O*-butyryl-2,6-di-*O*-pentyl)-γ-CD in OV 1701; 30% heptakis-(6-*O*-*t*-butyldimethylsilyl-2,3-di-*O*-methyl)-β-CD; 30% heptakis-(6-*O*-*t*-butyldimethylsilyl-2-*O*-methyl-3-*O*-pentyl)-β-CD in 40% OV 1701. It was established that chiral stationary phases containing modified CDs were suitable for the enantiomeric separation of chlorinated bis(propyl) ether pollutants in river water. The enantiomer ratios from the Elbe river (Table 2.3) indicated the enantioselective biodegradation of pollutants.

Other Organohalogen Analytes or PCB and Arylhalogen Analysis

Various CD stationary phases have also found application in the GC analysis of fluorinated alkylbenzenes. The solutes separated well on permethylated CDs and the ratio of phenyl groups in the achiral polysiloxanes exerted a beneficial effect on the separation.[41] Hydrophobic and van der Waals forces accounted for the interaction of solutes in the stationary phase. Perethylated, perbutylated and peroctylated β-CDs have been employed for the analysis of halobenzenes and other environmental pollutants.[42–44] The results indicated that more than one interactive force (van der Waals, hydrogen bonding, dipole–dipole and hydrophobic interactions) can be involved in the separation mechanism. The successful separation of the atropisomeric alkylated and polychlorinated biphenyls on CD derivatives has also been reported. The method used octakis-(6-*O*-methyl-2,3-di-*O*-pentyl)-γ-CD and octakis-(2,6-di-*O*-methyl-3-*O*-pentyl)-γ-CD as chiral selectors.[45] The same chiral selectors have been used for the separation of PCBs, polychlorinated and organophosphorus pesticides and pyrethroids.[46] The seasonal dependence of PCBs[56] has been also measured in blue mussels (*Mytilus edulis* L.) on a chiral stationary phase consisting of 50% heptakis-(6-*O*-*t*-

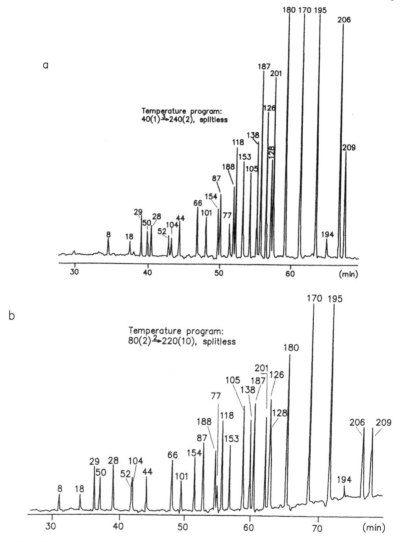

Figure 2.5 *HRGC/ECD of the PCB mixture SRM 2262 (NIST) on 12 m OV 1701 (a) and 13 m 10% permethylated β-cyclodextrin in OV 1701(b)*
(Reprinted with permission from ref. 38)

butyldimethylsilyl-2,3-di-*O*-methyl)-β-CD in OV 1701. PCBs have been detected in mussels and their concentration was established.[47,48] A column switching technique (achiral column followed with a chiral) has also been employed for the separation of PCBs and the method has been used for the measurement of PCBs in biological matrices such as human milk, doe liver and eel.[49,50] A similar multidimensional GC technique has been used for the separation and quantitative determination of PCBs in soils and sediments,[51,52] and for the determination of the

Table 2.3 *Enantiomeric proportions of chiral tetrachlorobis(propyl) ethers (BPE) in the Elbe river (EE = enantiomeric excess)*

| Location | No. | 1,3,2',3'-BPE | | 2,3,2',3'-BPE | | |
		(−)/(+)	e.e. (%)	(R,R')/(S,S')	e.e. (%)	M/(S, S')
February 1995						
Bilina		1.02	1.0	0.94	2.9	1.91
Schmilka	P1	0.85	7.8	0.63	22.5	1.66
Magdeburg	P3	0.92	3.8	1.05	2.5	2.06
Schnackenburg	P5	0.89	5.6	0.78	12.3	1.87
Zollenspieker	P6	1.01	0.5	0.94	2.9	1.92
Seemanshöft	P7	0.87	6.5	0.41	41.3	1.16
Grauerort	P8	1.05	2.5	1.02	1.0	2.01
Cuxhaven	P9	1.20	9.2	1.59	22.7	2.88
September 1995						
Schmilka	P1	0.86	7.4	0.54	29.8	0.74
Zehren	P2	1.00	0.0	0.83	9.1	0.83
Magdeburg	P3	0.84	8.7	0.53	30.8	1.75
Tangermünde	P4	0.82	9.9	0.44	38.5	1.42
Schnackenburg	P5	0.72	16.3	0.27	57.4	0.92

(Reprinted with permission from ref. 40)

enantiomeric ratio of chiral PCB 95, PCB 132 and PCB 149 in shark liver (*C. coelolepis*).[53] A Chirasil-Dex column has been applied for the study of the enantiomerisation barrier of atropisomeric 2,2', 3,3', 4,6'-hexachlorobiphenyl (PCB 132),[54] and for the determination of the elution order of (+)- and (−)-enantiomers of stable atropisomeric PCBs.[55] Another study used permethylated β-CD as chiral selector for the measurement of the interconversion energy barriers of atropisomers of some PCBs.[56] The separation capacity of seven commercially available chiral capillary columns was compared using PCB atropisomers as model compounds. It was established that the separation was highly dependent on the type of stationary phase and the atropisomers to be separated.[57]

A combined method using pyrenyl-silica HPLC and a Chirasil-Dex chiral capillary column has been developed for the separation of coplanar and chiral PCBs. The method has been employed for the measurement of PCBs in cow milk and otter.[58]

Other Analytes

Enantiomeric alkyl nitrates have been separated on a heptakis-(3-*O*-acetyl-2,6-di-*O*-pentyl)-β-CD (LIPODEX-D) stationary phase.[59] However, chiral alkyl nitrates in air could only be separated by coupling a chiral CG column to a polyalkylenglycol (PAG) column (Figure 2.6).

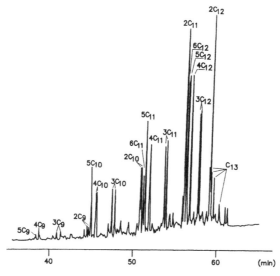

Figure 2.6 *Chiral separation of C_9-C_{12} alkyl nitrates present in a sample of rural air (from the vicinity of Ulm). Conditions: Chiral GC (LIPODEX + PAG)/ECD; temp. prog. 40 °C (2 min), rate A 2 °C min^{-1} to 140 °C (1 min), rate B 10 °C min^{-1} to 220 °C*
(Reprinted from ref. 59)

Mixed stationary phases containing heptakis-(2,3,6-tri-*O*-pentyl)-β-cyclodextrin, AgNO$_3$, TlNO$_3$, polyethylene glycol 400 (PEG 400) and bentone 34 have also been employed for the separation of xylene isomers using packed columns.[60] It was concluded that the mixed stationary phases show both positive and negative synergistic effects depending on the type of the phase and on the isomer pair to be separated. Synthesis and application of a polysiloxane stationary phase with a β-CD side chain has also been reported for the separation of naphthalene homologues (Figure 2.7).[61]

The fact that some enantiomers cannot be separated on this stationary phase suggests that the efficacy of chiral separation greatly depends on both the chemical structure of guest and host molecules.

Benzyl-β-CD derivatives have also been synthesised and their separation characteristics have been elucidated.[62] Capillary columns were coated with heptakis-(2,6-di-*O*-benzyl-3-*O*-pentyl)-β-CD, heptakis-(2,6-di-*O*-benzyl-3-*O*-methyl)-β-CD, and heptakis-(2,6-di-*O*-benzyl-3-*O*-acetyl)-β-CD. Pyrolysis GC indicated that the benzyl derivatives are stable up to 300 °C: therefore they can be used as GC stationary phases. The results proved that benzyl β-CD derivatives are novel and efficient stationary phases for the separation of some classes of positional isomers.

The influence of the chemical structure of polysiloxanes as carriers for modified CDs has been investigated in detail.[63] Side-chain crown ether polysiloxane (PDB-14-C4), side-chain liquid-crystalline polysiloxane (PSC-3), heptakis-

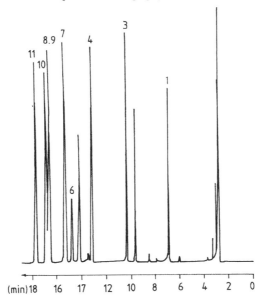

Figure 2.7 *Separation of naphthalene homologues on a capillary column (fused-silica capillary tubes (0.25 mm ID)) at 160 °C. Carrier gas (nitrogen) velocity 17.8 cm s⁻¹. Peaks: 1 = naphthalene; 2 = 2-methylnaphthalene; 3 = 1-methylnaphthalene; 4 = biphenyl; 5 = 2,6-dimethylnaphthalene; 6 = 1,7-dimethylnaphthalene; 7 = 1,6-dimethylnaphthalene; 8.9 = 1,4- and 2,3-dimethylnaphthalene (unresolved); 10 = 1,5-dimethylnaphthalene; 11 = 1,2-dimethylnaphthalene*
(Reprinted with permission from ref. 61)

(2,3,6-tri-*O*-methyl)-β-CD (PM β-CD) and heptakis-(2,3,6-tri-*O*-ethyl)-β-CD (PF β-CD) were synthesised, mixed in various ratios and applied as coatings in fused-silica capillaries. It was found that some isomers did not follow the additivity rule: their α value was higher or lower on the mixed stationary phase than on the stationary phases coated with pure components. This phenomenon was tentatively explained by the new term 'coordination effect'; however, the exact physicochemical meaning of the new term was not defined.

The search for new CD derivatives suitable for stationary phase in GC is more and more intensive. Thus, seven new modified β-CD derivatives were synthesised: heptakis-[2,6-di-*O*-benzyl-3-*O*-(4-nitrobenzyl)]-β-CD; heptakis-[2,6-di-*O*-pentyl-3-*O*-(5-hexenyl)]-β-CD; heptakis-[2,6-di-*O*-pentyl-3-*O*-trifluoroacetyl)-β-CD; heptakis-(2,3,6-tri-*O*-pentyl)-β-CD; heptakis-(2,6-di-*O*-cynnamyl-3-*O*-methyl)-β-CD; heptakis-(2,6-di-*O*-cynnamyl-3-*O*-acetyl)-β-CD and heptakis-(2,6-di-*O*-cynnamil-3-*O*-trifluoroacetyl)-β-CD; their separation properties were tested and compared.[64] The presence of double bond and aromatic ring structure [for example in the separation of dimethylphenol isomers (Figure 2.8)] influenced the separation characteristics. It was further assumed that induced dipole–dipole and π–π interactions exert a considerable effect on the retention.

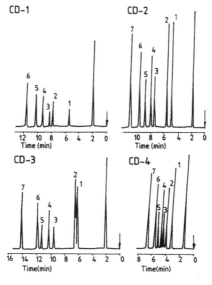

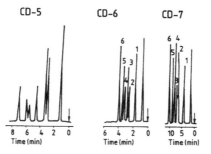

Figure 2.8 *Separation of dimethylphenol (DMP) isomers. CD-1 (150 °C) on list columns:*
1 = 2,6-DMP;   2 = 2,4-DMP;   3 = 2,5-DMP;   4 = 2,3-DMP;   5 = 3,5-DMP;
6 = 3,4-DMP. CD-2 (160 °C):   1 = 2,6-DMP;   2 = phenol;   3 = 2,4-DMP;
4 = 2,5-DMP;   5 = 2,3-DMP;   6 = 3,5-DMP;   7 = 3,4-DMP. CD-3 (150 °C):
1 = phenol;   2 = 2,6-DMP;   3 = 2,4-DMP;   4 = 2,5-DMP;   5 = 2,3-DMP;
6 = 3,5-DMP;   7 = 3,4-DMP. CD-4 (170 °C): 1 = phenol;   2 = 2,6-DMP;
3 = 2,4-DMP; 4 = 2,5-DMP; 5 = 2,3-DMP; 6 = 3,5-DMP; 7 = 3,4-DMP. CD-
5 (130 °C): 1 = 2,6-DMP; 2 = phenol; 3 = 2,4-DMP; 4 = 2,5-DMP; 5 = 2,3-
DMP; 6 = 3,5-DMP; 7 = 3,4-DMP. CD-6 (130 °C): 1 = 2,6-DMP + phenol;
2 = 2,4-DMP; 3 = 2,5-DMP; 4 = 2,3-DMP; 5 = 3,5-DMP; 6 = 3,4-DMP. CD-
7 (130 °C):   1 = 2,6-DMP;   2 = 2,4-DMP;   3 = 2,5-DMP;   4 = 2,3-DMP;
5 = 3,5-DMP; 6 = 3,4-DMP
(Reprinted with permission from ref. 64)

The retention properties and pyrolysis of heptakis-[3-*O*-methyl-2,6-di-*O*-(meth-oxybenzyl)]-β-CD (MMBCD) were also studied in detail.[65,66] It was established that MMBCD is stable up to 300 °C; therefore it can be employed as stationary phase in GC. It has been assumed that diastereomeric dipole/dipole and/or H-bond formation account for the chiral separation and not the inclusion complex formation.

The interaction of alkyl nitrates with heptakis-(3-*O*-acetyl-2,6-di-*O*-pentyl)-β-CD has been studied in detail. The selectivity of the interaction depended considerably on the temperature, and van der Waals interactive forces governed the formation of CD–analyte complexes.[67]

Separation of Positional and Optical Isomers of Pharmaceuticals

GC stationary phases containing CDs or CD derivatives have not been employed frequently in the analysis of pharmaceuticals. It may be due to the low volatility and high decomposition rate at elevated temperatures of many drug compounds. However, volatile compounds, such as anaesthetics, have been extensively investigated by GC methods. The thermodynamics of the enantiomeric separation of the chiral inhalation anaesthetics enflurane, isoflurane and desflurane has been studied in detail.[68] Fused silica capillaries coated with 5.23% octakis-(3-*O*-butanoyl-2,6-di-*O*-*n*-pentyl)-γ-CD and 94.77% polysiloxane SE-54 and with 10.135% octakis-(3-*O*-butanoyl-2,6-di-*O*-*n*-pentyl)-γ-CD were used and a 9.865% polysiloxane SE-54A column coated with SE-54 was employed as reference column. The retention and separation of enantiomers of enflurane (1), isoflurane (2) and desflurane (3) were very dependent on the oven temperature (Figure 2.9). Below the isoenantioselective temperatures (T_{iso}) the D-enantiomer was eluted after the L-enantiomer (enthalpy-controlled separation). Above T_{iso} the elution order was reversed (entropy-controlled separation). High differences were found

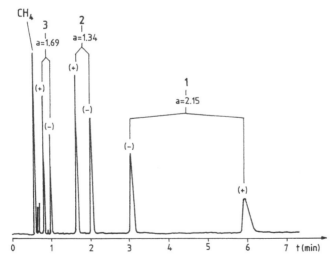

Figure 2.9 *Simultaneous analytical gas-chromatographic separation of the inhalation anesthetics enflurane (1), isoflurane (2) and desflurane (3). Fused-silica capillary column (25 m × 0.25 mm ID) coated with 10.135% octakis-(3-O-butanoyl-2,6-di-O-n-pentyl)-γ-CD dissolved in 9.865% polysiloxane SE-54, film thickness 0.5 μm, at 26 °C. Carrier gas: 1.1 bar helium*
(Reprinted with permission from ref. 68)

between the thermodynamic characteristics of the interaction of the anaesthetics reflecting differences in the capacity of these analytes to interact with the γ-CD derivatives. A preparative enantiomeric separation of the same compounds has also been carried out on a 1 m $\times$ 24 mm column packed with a mixture of SE-54 and octakis-(3-*O*-butanoyl-2,6-di-*O*-n-pentyl)-γ-CD on Chromosorb P AW DMCS,[69] which enabled the mode of action of the individual enantiomers to be studied. Another method used a gas chromatographic simulated moving bed (GC–SMB) unit for the continuous enantiomer separation of enflurane.[70] The temperature dependence of the separation of the fluoroether anaesthetics (desflurane, enflurane and isoflurane) has also been measured on nine CG-based enantioselective stationary phases[71] and the trifluoroacetyl CD column has been employed for the preparative scale separation of the enantiomers of isoflurane[72] and those of enflurane.[73–75]

Hexobarbital, methylphenobarbital and pentobarbital enantiomers have been separated under GC, SFC, HPLC and CE conditions using permethylated β-CD chemically bonded to dimethylpolysiloxane through an octamethylene spacer as a chiral selector.[76,77] The enantiomer separation of flumecinol and related compounds using a permethylated-β-CD chiral selector has also been reported.[78] The enantiomers of the mucoregulatory drug *trans*-sobrerol have been separated on a preparative scale on heptakis-(2,6-di-*O*-methyl-3-*O*-pentyl)-β-CD and the method was proposed as a general technique for the preparative isolation of enantiomers.[79] Trifluoroacetyl and acetyl derivatives of amphetamine and methamphetamine have been successfully separated on 2,6-di-*O*-pentyl-3-*O*-trifluoroacetyl-β-CD, 2,6-di-*O*-pentyl-3-*O*-trifluotoacetyl-γ-CD, and 2,6-di-*O*-pentyl-3-*O*-propionyl-γ-CD stationary phases.[80]

Separation of Positional and Optical Isomers of Aroma Compounds and Food Components

Early results in the chirospecific analysis of fragrance and flavour compounds[81] and in the application of CD derivatives in the GC separation of their enantiomers have been reviewed.[82,83] The advantages and disadvantages of the application of enantioselective GC in the purity control of essential oils have also been discussed.[84]

The separation capacity of various CD derivatives and that of their mixtures with achiral stationary phases has been vigorously discussed using synthetic aroma components as model compounds. The modified CDs permethylated-β-CD and (2,6-di-*O*-methyl-3-*O*-trifluoroacetyl)-β and γ-CDs were synthesised and diluted with Carbowax 20M.[85] In the majority of cases a good relationship was found between the calculated minimal energy of host–guest interaction and the GC elution order of the enantiomers of model aroma compounds. The irregular behaviour of some amino acids and lactones was explained by the possibility of hydrogen-bond formation and sterical orientation of the side chain. Thick-film wide-bore columns were employed for the micropreparative enantiomeric separation of γ-decalactone from a nature-identical mango aroma and other volatile compounds (Figure 2.10).[86] Because of the high separation power the method was proposed for the isolation of pure enantiomers from complex natural mixtures in the μg range. Similar thick-film wide-bore columns have been used for the semi-preparative enantiomeric separation of 22 enantiomeric pairs in the low mg range.[87] A marked effect of the type and concentration of the diluting apolar phase on the separation capacity of heptakis-(2,6-di-*O*-methyl-3-*O*-pentyl)-β-CD has been established.[88,89] Similar results were achieved with heptakis-(2,3-di-*O*-acetyl-6-*O*-*t*-butyldimethylsilyl)-β-CD.[89] Another study established that the enantioselectivity of *t*-butylmethyl-β-CD is superior to that of the permethyl derivative.[90] The enantioselectivity of 6-*t*-butyldimethylsilyl-α-CD, 6-*t*-butyl dimethylsilyl-2,3-diethyl and dipropyl-β-CD has been compared using racemic lactones. The data indicated that the selectivity depended not only on the type of chiral selector but also on the chemical structure of the enantiomers to be

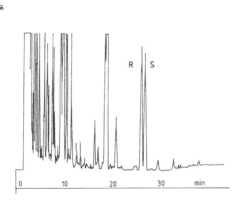

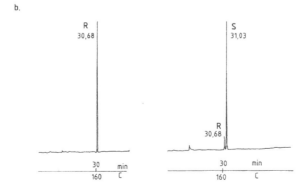

Figure 2.10 *Capillary GC patterns of the nature-identical mango aroma on the AcTBDMS-β-CD/PS-086 micropreparative column (a) and those of the resulting isolated enantiomers on the corresponding analytical column (b)* (Reprinted with permission from ref. 84)

separated.[91] The application of a polysiloxane-bonded and immobilized CD,[92] and that of 6[A],6[B]-β-CD-hexasiloxane copolymers[93] have also been reported. Acyclic derivatives of sugar enantiomers have also been separated by capillary GC using permethylated α-, β- and γ-CDs as chiral selectors.[94] Alditol trifluoroacetate and trifluoroacetylated aldose diethyl dithioacetal derivatives of the enantiomeric monosaccharides arbinose, fucose, galactose, glucose, lyxose, mannose, rhamnose, ribose, 2-deoxyribose and xylose were separated on columns of permethylated α-, β- and γ-CDs dissolved in 80% poly-(35% diphenyl-65% dimethyl-siloxane) (Table 2.4). The results indicated that not each enantiomer pair can be resolved on permethylated CD stationary phases. However, trifluoroacetylated alditols were better separated than the trifluoroacetylated aldose diethyl dithioacetals.

Enantiomer separation of monoterpenes and terpenoids has been widely studied. Thus, monoterpene enantiomers in the needle oils of Scots pine (*Pinus sylvestris* L.) and juniper (*Juniperus communis* L.) have been measured using permethylated β-CD as a chiral selector and for the elucidation of biogenetic relationships.[95,96] The same method has been applied for the enantiomeric analysis of monoterpene hydrocarbons in xylem and needles of *Picea abies*.[97] Enantiomers of monoterpene, sesquiterpene and diterpene hydrocarbons have

Table 2.4 *GLC data for trifluoroacetylated alditols of sugar enantiomers on three chiral columns: α-Dex 120, β-Dex 120 and γ-Dex 120*

| | α-Dex 120 | | β-Dex 120 | | γ-Dex 120 | |
Parent sugar	$t_R{}^a$	k^b	$t_R{}^a$	k^b	$t_R{}^a$	k^b
Ara	0.71	1.08	0.73	1.09	0.67	1.05
Fuc	0.51	1.00	0.47	1.00	0.42	1.00
Gal	1.00	–[c]	1.00	–	1.00	–
Glc	0.86	1.01	0.86	1.05[*d]	0.81	1.06[*]
Lyx	0.71	1.08	0.73	1.09	0.67	1.05
Man	0.65	1.00	0.63	1.00	0.60	1.04[*]
Rha	0.45	1.00	0.43	1.04	0.40	1.00
Rib	0.57	–	0.47	–	0.50	–
2-Deoxy-rib	1.12	1.02	1.24	1.00	1.08	1.00
Xyl	0.84	–	0.78	–	0.85	–
Abe/Col	1.22	1.04	1.35	1.05	1.08	1.07
Tyv[c]	1.06	–	1.14	–	0.94	–

[a] t_R, Retention time for the D-form derivative relative to that of trifluoroacetylated galacticol, which eluted at about 10 min (α-Dex), 12 min (β-Dex) and 14 min (γ-Dex).
[b] k, Separation factor, t'_{R2}/t'_{R1}, where $t'_{R2} > t'_{R1}$ and are the corrected retention times for the slower and the faster component (D- or L-derivative, see below). If $k < 1.005$ it is given as 1.00. [c]–, Not applicable, the relative retention time for comparison. [d] *, The D-form eluted before the L-form for all derivatives except for those marked with an * or which had k-value 1.00. [e]The L-form (3,6-dideoxy-L-mannose, ascarylose) was not available.
(Reprinted with permission from ref. 91)

been measured in the resin of *Pinus pseudostrobus* and in the extract of *Pellia epiphylla* using heptakis-(6-*O*-*t*-butyldimethylsilyl-2,3-*O*-methyl-β-CD as chiral selector.[98] The enantiomer composition of the volatiles in the needle oil of *Pinus peuce* Griseb. has been elucidated by using a heptakis-(2,3,6-tri-*O*-methyl)-β-CD/SPB-35 stationary phase.[99] Oxygenated monoterpenoids have been isolated and identified in the petals of *Rosa damascena* Mills by multilayer coil countercurrent chromatography and multidimensional gas chromatography using 2[6-(*t*-butyldimethylsilyl)]-2,3-dimethyl-β-CD dissolved in SE 54 as chiral selector.[100]

Enantiomers of monoterpene hydrocarbons and monoterpene alcohols of mandarin oils have been separated and quantitatively determined by multidimensional capillary GC.[101] The efficacy of 12 CD derivatives was compared with that of Chirasil-Val for the enantiomeric separation of bicyclic terpenes, sesquiterpenes and aromatics. The results indicated that the elution order may be different on different CD stationary phases.[102] CD derivatives have found application in the enantiomer separation of sesquiterpene hydrocarbons from various essential oils such as *Piper nigrum*, *Scapania undulata*, *Barbilophozia barbata*, *Scapania aequiloba*, *Scapania nemorea*, *Preissia quadrata*, *Conocephalum conicum*, *Frullania tamarisci*, *Copaiba balsam*, *Piper cubeba* and Orange.[103] The enantiomers of sesquiterpene hydrocarbons from the essential oil of *Cedrela odorata* L. have been separated by two-dimensional GC. Chiral capillary column was coated with 20% heptakis-(2,6-di-*O*-methyl-3-*O*-pentyl)-β-CD dissolved in OV 1701.[104] A similar method has been employed for the identification and chiral separation of sesquiterpene hydrocarbons in the essential oil of *Meum athamanticum*,[105] *Lophocolea bidentata*,[106] *Preissia quadrata*,[107] *Dumortiera hirsuta*,[108] *Lophocolea heterophylla*,[109] new sesquiterpene of *Lepidozia reptans*[110] and *Marchantia polymorpha* ssp. *aquatica*.[111] (+)-β-Caryophyllene has also been identified and separated from specimens of *Pellia endiviifolia* and *P. epiphylla* by the same method.[112] (+)-(3*S*,4*S*)-3-Butyl-4-vinylcyclopentene has been identified in the essential oils of the marine brown algae *Dictyopteris prolifera* by using chiral GC with CP-Cyclodex 236 M column.[113] A permethylated β-CD chiral stationary phase has been applied for the study of the biotransformation of linalool to furanoidand pyranoid linalool oxides by *Aspergillus niger*.[114]

Enantiomeric GC carried out on a 50% heptakis-(2,6-di-*O*-methyl-3-*O*-pentyl)-β-CD in OV 1701 stationary phase helped the identification of new components in the essential oil of the liverwort *Mannia fragrans*.[115] The occurrence of cadina-3,5-diene and δ-amorphene in manuka oil (*Leptospermum scoparium*) and in the chemotype of the liverwort *Conocephalum conicum* has been proven by enantioselective GC using a heptakis-(6-*O*-*t*-butyldimethylsilyl-2,3-di-*O*-methyl)-β-CD dissolved in 50% OV 1701.[116] The same chiral stationary phase was employed for the identification of the new eudesmane sesquiterpene (−)-eudesma-1,4(15),11-triene from the essential oil of *Callitris intratropica*.[117] The direct enantiomeric separation of limonene and linalool from essential oils such as rose, myrcia, lemon-grass and orange oils has been achieved by using a dual column system and per-*n*-pentyl-β-CD and hexakis-(3-*O*-trifluoroacetyl-2,6-di-*O*-pentyl)-α-CD as chiral selectors.[118,119] GC coupled with isotope ratio MS has been used for the authenticity control of the essential oil from *Coriandrum sativum* L. Enantiomer separations were carried out on a heptakis-(2,3-di-*O*-methyl-6-*O*-*t*-butyldimethyl-silyl)-*β*-CD/PS 268 stationary phase.[120] The same chiral stationary phase has been used for the study of the biogenesis of geranium oil components.[121] Characteristic flavour constituents of mandarin essential oil have also been investigated by GC-isotope ratio MS using heptakis-(2,3-di-*O*-methyl-6-*O*-*t*-butyldimethylsilyl)-*β*-CD/OV 1701 (45:55%) as chiral selector.[122] The chiral separation of (*E,Z*)-2,3-dihydrofarnesals was carried out on a heptakis-(2,3-di-*O*-acetyl-6-*O*-*t*-butyldi-methylsilyl)-*β*-CD/PS 268 stationary phase. It was found that the enantiomeric ratio is in favour of the (3*S*) enantiomer in the scent of orchids (*Aerides jarckianum*) and in the blossom fragrance of *Citrus limon*.[123]

Not only the enantiomers of the constituents of essential oils but also the aroma components of foods have been extensively investigated. The aroma compounds of raspberry cultivars were separated on a capillary column coated with 10% octakis-(2,6-dimethyl-3-fluoroacetyl)-*γ*-CD dissolved in 90% OV 1701.[124] The evaluation of the data by principal component analysis made possible the differentiation of cultivars. The composition of volatile components of Andes berry (*Rubus glaucus* Benth) have been separated and quantitated by multidimensional GC, the second column being coated with heptakis-(2,3-diacetyl-6-*O*-*t*-butylmethylsilyl)-*β*-CD/OV 1701. The retention time and concentration of 44 volatile constituents have been determined.[125] Multidimensional GC with sulfur chemiluminescence detection has been employed for the study of the enantiomeric distribution of 2-pentanethiol in guava fruit (*Psidium guajava* L.) using 30% heptakis-(2,3-di-*O*-ethyl-6-*O*-*t*-butyldimethylsilyl)-*β*-CD as chiral selector.[126] It was concluded that the method may help the authenticity test of flavourings.

Chiral sulfur-containing alcohols have been determined in yellow and purple passion fruits, the enantiomers being separated on a similar stationary phase. The procedure has been suggested as a control method of authenticity.[127] Another multidimensional GC technique has been employed for the separation of aliphatic *β*-D-glucosides from fruits of *Carica pubescens*. Both heptakis-(2,6-di-*O*-methyl-3-*O*-pentyl) and heptakis-(2,3-di-*O*-acetyl-6-*O*-*t*-butyldimethylsilyl)-*β*-CD mixed with OV 1701 were employed as stationary phases.[128] Enantiomers of ethyl acetate from wines were separated on a capillary column coated with 30% heptakis-(2,3-di-*O*-methyl-6-*O*-*t*-butyldimethylsilyl)-*β*-CD dissolved in SE 52. It was stated that the method may contribute to the quality assurance of wine.[129] The absolute configuration of 1,3-dioxanes has been determined in French cider using HRGC–MS, and heptakis-(2,3-di-*O*-acetyl-6-*O*-*t*-butyldimethylsilyl)-*β*-CD dissolved in OV 1701 served as chiral selector.[130] Enantiomers of solerone, solerol and other *α*-ketols were measured in sherry by GC.[131] Enantiomers of butylhexahydrophthalide in the extract of celeriac (*Apium graveolens* L. var. *rapaceum*) have been separated on heptakis-(2,3-di-*O*-acetyl-6-*O*-*t*-butyldimethylsilyl)-*β*-CD stationary phase. It was found that all four stereoisomers of (3a,7a)-*cis*-3-butylhexahy-drophthalide are present in the extract.[132] Similar methods were employed for the analysis of the enantiomers of 3-butylphthalide and 3-butylhexahydrophthalide in celery, celeriac, celery seed and fennel extracts.[133,134] The same chiral selector has been used for the enantioselective analysis of $\delta(\gamma)$-lactones C_8–C_{18} in butter, cream, evaporated milk, cheese, margarine, soya milk and coconut.[135] In order to elucidate the presence of synthetic menthol, menthol isomers have been separated in some commercial products using a CD capillary column and MS detection.[136] The enantiomeric distribution of limonene and limonene-1,2-epoxide in lemon peel has been elucidated by multidimensional GC using FID and SIM-MS as detection methods. Chiral separations were performed on a Chirasil-γ-Dex column.[137] Isomers of limonene, carvone and carveols were analysed in seeds of dill (*Anethum graveolens* L.) and annual (*Carum carvi* L.f. *annuum*) and biennial caraway (*Carum carvi* L.) varieties. Enantiomer separations were performed on an octakis-(6-*O*-methyl-2,3-di-*O*-pentyl)-γ-CD (80% in OV 1701) chiral stationary phase.[138] Enantioselective analysis of dill ether and its *cis*-stereoisomers has been carried out on chiral stationary phase.[139] Isomers of the decomposition products of the aroma precursors C_{13} norisoprenoid glycosides from stinging nettle (*Urtica dioica* L.) have been identified by GC–MS and FTIR. Enantiomeric separation was carried out on a 2,6-dimethyl-3-pentyl-β-CD/OV 1701 stationary phase.[140] Permethylated CD chiral selectors have been employed for the enantiomeric separation of α-campholene and fencholene derivatives using single-[141] and double-column techniques.[142] Preparative enantiomer separation of 1-octenyl-3-acetate, a mushroom odour aroma compound, has been performed on dual columns, modified β-CD being the chiral selector.[143,144] The isomers of the tobacco flavour compound 2,3,5,6,8a-hexahydro-2,5,5,8a-tetramethyl-7*H*-1-benzo-pyran-7-one (3,4-dihydro-3-oxoedulans) have been separated by multidimensional GC using octakis-(2,6-di-*O*-methyl-3-*O*-pentyl)-γ-CD as chiral selector.[145] Enantiomers of 3-methylthiobutanal have been separated on a chiral stationary phase containing octakis-(2,3-di-*O*-butyryl-6-*O*-*t*-butyldimethylsilyl)-γ-CD. Olfactometry indicated that only the (*R*) form shows the characteristic odour of cooked potatoes; the other is odourless.[146]

γ-Thiolactones [5-alkyldihydro-2(3*H*)-thiophenones] and δ-thiolactones (6-alkyltetrahydro-2*H*-thiopyran-2-ones) were enantioseparated on a heptakis-(2,3-di-*O*-acetyl-6-*O*-*t*-butyldimethylsilyl)-β-CD stationary phase and the odour thresholds were measured. It was found that the odour threshold depends on both the ring size and chain length.[147] The application of CDs in the authenticity control of natural flavours and fragrances has been previously reviewed.[148] The floral fragrance compounds from the early flowering shrub *Daphne mezereum* were separated on a permethylated-β-CD column.[149] The enantiomeric separation factors of a considerable number of compounds have been collected for stationary phases containing heptakis-(2,3-di-*O*-methyl-6-*O*-*t*-butyldimethylsilyl)-β-CD[150] and heptakis-(2,3-di-*O*-acetyl-6-*O*-*t*-butyldimethylsilyl)-β-CD.[151]

Molecular modelling has been successfully employed for the study of the interaction mechanism of permethylated CDs with guest molecules in the gas phase.[152] Enantiomers of α-pinene were separated only on the chiral stationary phases containing permethylated β-CD (Figure 2.11). Calculations indicated that in the case of α-pinene the weak electrostatic interactions between the double bonds and the glucosidic oxygen atoms of permethylated β-CD may account for the enantiomeric separation.

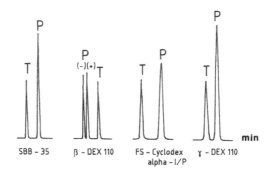

Figure 2.11 *Separation of tricyclene (T), (1S)-(−)-α-pinene and (1R)-(+)-α-pinene (P) on permethylated cyclodextrin phases dissolved in moderately polar polysiloxanes and elution order on the non-chiral polysiloxane phase SPB 35* (Reprinted with permission from ref. 152)

Separation of Positional and Optical Isomers of Miscellaneous Organic Compounds

CD stationary phases have found application in the determination of enantiomer ratio of bioactive compounds even in complicated matrices. Thus, the enantiomeric separation of the plant growth regulator 12-oxophytodienoic acid (12-oxo-PDA) extracted from plant tissues has been reported.[153] Enantiomeric separation of the *trans-* and *cis*-12-oxo-PDA methyl esters are shown in Figure 2.12. It was established that permethylated β-CD separated *trans*-isomers well but was not suitable for the separation of *cis*-isomers, which can be achieved on a permethylated γ-CD stationary phase. Interestingly, only *cis*-(+)-12-oxo-PDA was detected in the leaves of both *Hordeum vulgare* and *Arabidopsis thaliana*.

Enantiomers of methyl, pentyl, acetyl and trifluoroacetyl derivatives of low-boiling secondary alcohols have been separated on a chiral stationary phase consisting of 50% heptakis-(6-O-t-butyldimethylsilyl-2,3-di-O-acetyl)-β-CD and 50% OV 1701. It was found that the total molecular asymmetry governs the enantiomer separation.[160–163,154–157] Alcohols, diols and phenyl-substituted amines have also been successfully separated on a trifluoroacetyl-CD phase.[158,159] The retention characteristics of various mixtures of heptakis-(2,3,6-tri-O-pentyl)-β-CD and OV 1701 have been measured using miscellaneous analytes as model compounds. The data suggested the interaction of perpentylated-β-CD with the polysiloxane.[160]

Another theoretical study established that the enantioselectivity of chiral stationary phases does not depend on their Rohrschneider-McReynold constants.[161] Enantiomer separation of cycloalkanols has been achieved on both permethylated α- and β-CD phases and the enthalpies and entropies of the interaction were calculated. It was found that the thermodynamic data are not correlated with the retention indices.[162] A set of 2,2-dialkyl-4-alkoxycarbonyl-1,3-dioxolane derivatives has been separated on 2,3-di-O-acetyl-6-O-t-butyldimethylsilyl-β and -γ-CDs. It was found that the geometric factors exert the highest

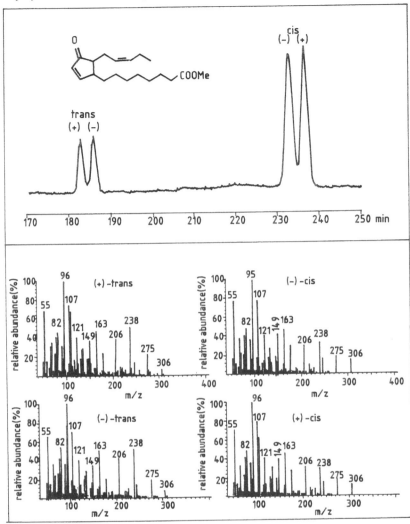

Figure 2.12 *Enantiomer separation of the trans and cis-12-oxo-phytodienoic acid methyl esters on a permethyl-γ-cyclodextrin capillary column at 160 °C and mass spectra of the four separation products. All mass spectra are averaged and background-subtracted. (Column 30 × 0.25 mm ID, coated with 0.2 μm film of 0.25% v/v cyclodextrin derivatives)*
(Reprinted with permission from ref. 153)

influence on the selectivity factors.[163] The retention characteristics of 6-*t*-butyldimethylsilyl-2,3-dimethyl-α-, β-, and γ-CD have been compared using various chiral model compounds. The advantageous parameters of the β-CD derivative has been proven.[164,165]

Some Novel Stationary Phase Materials

Novel heptakis(6-*O*-alkyldimethylsilyl-2,3-di-*O*-ethyl)-β-CD derivatives were synthesized and their separation capacity was compared using various racemic lactones as model compounds.[166] The results were compared with those obtained with a similar column coated with heptakis(6-*O*-*t*-butyldimethylsilyl-2,3-di-*O*-ethyl)-β-CD (TBDE-β-CD) in OV-1701 (Figure 2.13). The lower separation capacity of IPDE-β-CD and CHDE-β-CD was explained by the weak hydrophobicity of isopropyl group in PDE-β-CD and by the bulkiness of cyclohexyl group in CHDE-β-CD. The order of enantioselectivity on CD derivatives was: TXDE-β-CD $\geqslant$ TBDE-β-CD > IPDE-β-CD > CHDE-β-CD. The results proved again that the hydrophobicity of substituents and the steric hindrances formed by them are the decisive factors in the determination of enantioselectivity.

Other CD derivatives have also been tested as chiral. Thus, the synthesis of hexakis-(6-*O*-*t*-butyldimethylsilyl-2,3-di-*O*-ethyl)-α-CD (TBDE-α-CD), heptakis-(6-*O*-*t*-butyldimethylsilyl-2,3-di-*O*-ethyl)-β-CD (TBDE-β-CD), and heptakis-(6-*O*-tert-butyldimethylsilyl-2,3-di-*O*-propyl)-β-CD (TBDP-β-CD) was described.[167] It was concluded that the smaller cavity of α-CD makes it suitable for the enantiomeric separation of small molecules such as secondary alcohols and nitro-substituted alcohols. The enantioselectivity of heptakis(6-*O*-alkyldimethylsilyl-2,3-di-*O*-methyl)-β-CDs has also been studied in detail.[168]

The separation factors obtained on different chiral stationary phases are compiled in Table 2.5. The difference in the selectivities of β-CD derivatives was explained again by the different hydrophobicity of substituents and the steric correspondence between the guest molecule and the CD cavity.

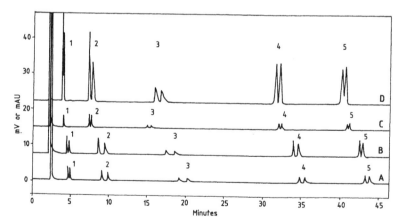

Figure 2.13 *Comparison of TBDE-β-CD (A), TXDE-β-CD (B), HDE-β-CD (C), and IPDE-β-CD (D) in OV-1701 dimensions % etc. as chiral selectors for the separation of enantiomers of racemic lactones, β-butyrolactone (1), α-methyl-γ-butyrolactone (2), γ-pantolactone (3), γ-nonanoic lactone (4), and γ-decanolactone (5). Oven temperature: programmed from 100 °C (held for 1 min) to 160 °C at 1.2 °C min⁻¹*
(Reprinted with permission from ref. 166)

Table 2.5 *Concentrations and enantiomeric ratios of bromocyclen in fish samples from the river Stör*

Sample number	Fish species	Sample station	Conc. $mg\,kg^{-1}$ fat	t_1/t_2 ER $(-)\,(+)$
F1	trout	Neumünster	0.21	0.66
F2	trout	Neuminster	0.11	0.71
F3	bream	Feldhusen	0.14	1.19
F4	bream	Feldhusen	0.24	1.22
F5	bream	Wittenbergen	0.06	1.08
F6	bream	Wittenbergen	0.05	1.34
F7	bream	Borsfleth	0.01	1.56
F8	bream	Borsfleth	0.01	2.13
F9	bream	Borsfleth	0.01	1.93
F10	bream	Borsfleth	0.02	2.05

ER: Enantiometric ratio, $(-)/(+)$-enantiomer of bromocyclen (reprinted with permission from ref. 8)

Crown ether capped β-CD has also been synthesised. Its separation characteristics for both optical and positional isomers were studied in detail and the results were compared with those obtained on other stationary phases containing β-CD derivatives.[169] The good separation characteristics of the new chiral phase was explained by the supposition that both the β-CD and crown ether cavities represent recognition sites for analytes and contribute to the enantiomeric discrimination.

2,3-Di-*O*-ethyl-6-*O*-*t*-butyldimethylsilyl-β and γ-CDs (ETTBS-β and γ-CDs) were also synthesised. Their enantioselectivities were measured and compared with those of 2,3-Di-*O*-methyl-6-*O*-*t*-butyldimethylsilyl-β- and γ-CDs (METBS-β- and γ-CDs).[170] It was stated that the new chiral stationary phases can be successfully employed for the separation of a wide variety of enantiomers. The type of nonchiral diluting agent influenced the enantioselectivity of ETTBS-γ-CD but exerted a negligible effect on ETTBS-β-CD. The synergistic effect of heptakis(2,3,6-tri-*O*-pentyl)-β-CD (HP-β-CD) and the liquid crystal *o*-methyl-*p*-phenylene-bis-(*p*-heptoxybenzoate) (MPBHpB) mixtures has been observed and exploited in the separation of positional isomers and other volatile compounds.[171] It was concluded that the synergistic effect depends on the composition of the stationary phase, the linear velocity of carrier gas and the chemical structure of the analyte pair to be separated.

References

1. P. Cardinael, E. Ndzie, S. Petit, G. Coquerel, Y. Combret and J. C. *J. High Resolut. Chromatogr.*, 1997, **20**, 560–564.
2. K. B. Lipkowitz, G. Pearl, B. Coner and M. A. Peterson, *J. Am. Chem. Soc.*, 1997, **119**, 600–610.
3. W. Vetter and V. Schurig, *J. Chromatogr. A*, 1997, **774**, 143–175.

4. G. Koske, G. Leupold and H. Parlar, *Fresenius' Envir. Bull.*, 1997, **6**, 489–493.
5. M. D. Müller, H.-R. Buser and C. Rappe, *Chemosphere*, 1997, **34**, 2407–2417.
6. V. Vetter, U. Klobes, B. Luckas and G. Hottinger, *Chromatographia*, 1997, **45**, 255–262.
7. H.-J. de Geus, R. Baycan-Keller, M. Oehme, J. de Boer and U. A. T. Brinkman, *J. High Resolut. Chromatogr.*, 1998, **21**, 39–46.
8. R. Baycan-Keller and M. Oehme, *J. High Resolut. Chromatogr.*, 1998, **21**, 298–302.
9. B. Bethan, K. Bester, H. Hühnerfuss and G. Rimkus, *Chemosphere*, 1997, **34**, 2271–2280.
10. W. Vetter, U. Klobes, K. Hummert and B. Luckas, *J. High Resolut. Chromatogr.*, 1997, **20**, 85–93.
11. M. Oehme, L. Müller and H. Karlsson, *J. Chromatogr. A*, 1997, **775**, 275–285.
12. R. Kallenborn, M. Oehme, W. Vetter and H. Parlar, *Chemosphere*, 1994, **28**, 89–98.
13. H.-R. Buser and M. D. Müller, *J. Agric. Food Chem.*, 1994, **42**, 393–400.
14. H.-R. Buser and M. D. Müller, *Environ. Sci. Technol.*, 1994, **28**, 119–128.
15. L. Alder, R. Palavinskas and P. Andrews, *Organohal. Comp.*, 1996, **28**, 410–415.
16. H. Karlsson, M. Oehme and L. Müller, *Organohal. Comp.*, 1996, **28**, 405–409.
17. R. Baycan-Keller and M. Oehme, *Organohal. Comp.*, 1997, **33**, 1–6.
18. H.-R. Buser and M. Müller, *Environ. Sci. Technol.*, 1993, **27**, 1211–1220.
19. W. A. König, I. H. Hardt, B. Gehrcke, D. H. Hochmuth, H. Hühnerfuss, B. Pfaffenberger and G. Rimkus, *Angew. Chem., Int. Ed. Engl.*, 1994, **33**, 2085–2087.
20. S. Mössner and K. Ballschmitter, *Fresenius' J. Anal. Chem.*, 1994, **348**, 583–589.
21. W. Vetter and B. Luckas, *J. High. Resolut. Chromatogr.*, 1995, **18**, 643–646.
22. R. L. Falconer, T. F. Biddleman, D. J. Gregor, R. Semkin and C. Teixeira, *Environ. Sci. Technol.*, 1995, **29**, 1297–1302.
23. L. M. Jantunen and T. F. Bidleman, *Organohal. Comp.*, 1995, **24**, 425–428.
24. R. L. Falconer, T. F. Bidleman and D. J. Gregor, *Sci. Total Environ.*, 1995, **160/161**, 65–74.
25. H.-R. Buser and M. D. Müller, *Environ. Sci. Technol.*, 1995, **29**, 664–672.
26. H. Hühnerfuss, J. Faller, R. Kallenborn, W. A. König, P. Ludwig, B. Pfaffenberger, M. Oehme and G. Rimkus, *Chirality*, 1993, **5**, 393–399.
27. S. Mössner, T. R. Spraker, P. R. Becker and K. Ballschmiter, *Chemosphere*, 1992, **24**, 1171–1180.
28. K. Möller, C. Bretzke, H. Hühnerfuss, R. Kallenborn, J. N. Kinkel, J. Kopf and G. Rimkus, *Angew. Chem., Int. Ed. Engl.*, 1994, **33**, 882–884.
29. W. Vetter, K. Hummert, B. Luckas and K. Skirnisson, *Sci. Total Environ.*, 1995, **170**, 159–164.
30. K. Hummert, W. Vetter and B. Luckas, *Chemosphere*, 1995, **31**, 3489–3500.

31. H. Hühnerfuss, B. Pfaffenberger and G. Rimkus, *Organohal. Comp.*, 1996, **29**, 88–93.

32. U. Müller, W. Vetter, K. Hummert and B. Luckas, *Organohal. Comp.*, 1996, **29**, 118–121.

33. S. Tanabe, P. Kumaran, H. Iwata, R. Tatsukawa and N. Miyazaki, *Mar. Pollut. Bull.*, 1996, **32**, 27–31.

34. B. Pfaffenberger, I. Hardt, H. Hühnerfuss, W. A. König, G. Rimkus, A. Glausch, V. Schurig and J. Hahn, *Chemosphere*, 1994, **29**, 1543–1554.

35. M. Oehme, R. Kallenborn, K. Wiberg and C. Rappe, *J. High. Resolut. Chromatogr.*, 1994, **17**, 583–588.

36. B. Pfaffenberger, H. Hühnerfuss, B. Gehrcke, I. Hardt, W. A. König and G. Rimkus, *Chemosphere*, 1994, **29**, 1385–1391.

37. M. Schneider and K. Ballschmiter, *Fresenius' J. Anal. Chem.*, 1995, **352**, 756–762.

38. B. Märker and K. Ballschmiter, *Fresenius' J. Anal. Chem.*, **356**, 98–99.

39. W. Vetter, U. Klobes, B. Luckas and G. Hottinger, *J. Chromatogr. A*, 1997, **769**, 247–252.

40. S. Franke, C. Meyer, M. Specht, W. A. König and W. Francke, *J. High Resolut. Chromatogr.*, 1998, **21**, 113–120.

41. R. Reinhardt, W. Engewald, O, Goj and G. Haufe, *Chromatographia*, 1994, **39**, 192–199.

42. P. Jing, R.-N. Fu, R.-J. Dai and J.-L. Gu, *Chromatographia*, 1996, **43**, 628–632.

43. D.-Q. Xiao, Y. Ling, Y.-X. Wen, R.-N. Fu, J.-L. Gu, R.-J. Dai and A.-Q. Luo, *Chromatographia*, 1997, **46**, 177–182.

44. V. Schurig and A. Glausch, *Naturwissenschaften*, 1993, **80**, 468–469.

45. W. A. König, B. Gehrcke, T. Runge and C. Wolf, *J. High Resolut. Chromatogr.*, 1993, **16**, 376–378.

46. I. H. Hardt, C. Wolf, B. Gehrcke, D. H. Hochmut, B. Pfaffenberger, H. Hühnerfuss and W. A. König, *J. High. Resolut. Chromatogr.*, 1994, **17**, 859–864.

47. H. Hühnerfuss, B. Pfaffenberger, B. Gehrcke, L. Karbe, W. A. König and O. Landgraff, *Organohal. Comp.*, 1994, **21**, 15–20.

48. H. Hühnerfuss, B. Pfaffenberger, B. Gehrcke, L. Karbe, W. A. König and O. Landgraff, *Marine Pollut. Bull.*, 1995, **30**, 332–340.

49. A. Glausch, G. J. Nicholson, M. Fluck and V. Schurig, *J. High Res. Chromatogr.*, 1994, **17**, 347–349.

50. A. Glausch, J. Hahn and V. Schurig, *Chemosphere*, 1995, **30**, 2079–2085.

51. E. Benická, R. Novakovsky, J. Hrouzek, J. Krupcik, P. Sandra and J. de Zeeuw, *J. High Resolut. Chromatogr.*, 1996, **19**, 95–98.

52. A. Glausch, G. P. Blanch and V. Schurig, *J. Chromatogr. A*, 1996, **723**, 399–404.

53. G. P. Blanch, A. Glausch, V. Schurig, R. Serrano and M. J. González, *J. High Resolut. Chromatogr.*, 1996, **19**, 392–396.

54. V. Schurig, A. Glausch and M. Fluck, *Tetrahedron: Asymmetry*, 1995, **6**, 2161–2164.

55. P. Haglund and K. Wiberg, *J. High Resolut. Chromatogr.*, 1996, **19**, 373–376.

56. J. Krupcik, M. Májeková, P. Májek, J. Hrouzek, E. Benická, F. Onuska, P. Sandra and J. de Zeeuw, *Fresenius' J. Anal. Chem.*, 1955, **352**, 696–698.

57. C. S. Wong and A. W. Garrison, *J. Chromatogr. A*, 2000, **866**, 213–220.

58. L. Ramos, B. Jiménez, M. Fernández, L. Hernández and M. J. González, *Organohal. Comp.*, 1996, **27**, 376–381.

59. M. Schneider and K. Ballschmitter. *Chem. Eur. J.*, 1996, **2**, 539–544.

60. L.-M. Yuan, R.-N. Fu, S.-H. Gui, X.-T. Xie, R.-J. Dai, X.-X. Chen and Q.-H. Xu, *Chromatographia*, 1997, **46**, 291–294.

61. H. B. Zhang, Y. Ling, R. N. Fu, Y. X. Wen and J. L. Gu, *Chromatographia*, 1997, **46**, 40–48.

62. D.-Q. Xiao, B. Q. Che, R. N. Fu, J. L. Gu, Y. X. Wen, Y. Ling and H. B. Zhang, *Chromatographia*, 1997, **44**, 393–398.

63. P. Jing, R.-N. Fu, R.-J. Dai, J.-L. Gu, Z. Huang and Y. Chen, *Chromatographia*, 1996, **43**, 546–550.

64. D.-Q. Xiao, Y. Ling, R.-N. Fu, J.-L. Gu, Z.-T. Zhao, R.-J. Dai, B.-Q. Che and A.-Q. Luo, *Chromatographia*, 1998, **47**, 557–564.

65. R. Dai, L. Ye, A. Luo, R. Fu, S. Zhang, G. Xie and S. Jin, *J. Anal. Appl. Pyrolysis*, 1997, **42**, 9–19.

66. H. Grosenick, V. Schurig, J. Constante and A. Collet, *Tetrahedron: Asymmetry*, 1995, **6**, 87–88.

67. M. Schneider and K. Ballschmiter, *J. Chromatogr. A*, 1999, **852**, 525–534.

68. V. Schurig and M. Juza, *J. Chromatogr. A*, 1997, **757**, 119–135.

69. M. Juza, E. Braun and V. Schurig, *J. Chromatogr. A*, 1997, **769**, 119–127.

70. M. Juza, O. di Giovanni, G. Biressi, V. Schurig, M. Mazzotti and M. Morbidelli, *J. Chromatogr. A*, 1998, **813**, 333–347.

71. A. Shitangkoon, D. U. Staerk and G. Vigh, *J. Chromatogr. A*, 1993, **657**, 387–394.

72. D. U. Staerk, A. Shitangkoon and G. Vigh, *J. Chromatogr. A*, 1994, **663**, 79–85.

73. D. U. Staerk, A. Shitangkoon and G. Vigh, *J. Chromatogr. A*, 1994, **677**, 133–140.

74. V. Schurig and H. Grosenick, *J. Chromatogr. A*, 1994, **666**, 617–625.

75. V. Schurig, M. Juza, B. S. Green, J. Horakh and A. Simon, *Angew. Chem., Ed. Engl.*, 1996, **35**, 1680–1682.

76. V. Schurig, M. Jung, S. Mayer, S. Negura, M. Fluck and H. Jakubetz, *Angew. Chem., Int. Ed. Engl.*, 1994, **33**, 2222–2223.

77. V. Schurig, M. Jung, S. Mayer, M. Fluck, S. Negura and H. Jakubetz, *J. Chromatogr. A*, 1995, **694**, 119–128.

78. R. Reinhardt, W. Engewald and S. Görög, *J. High Resolut. Chromatogr.*, 1995, **18**, 259–262.

79. I. Hardt and W. A. König, *J. Chromatogr. A*, 1994, **666**, 611–615.

80. H. L. Jin and T. E. Beesley, *Chromatographia*, 1994, **38**, 595–598.

81. P. Werkhoff, S. Brennecke, W. Breitschneider, M. Güntert, R. Hopp and H. Surburg, *Z. Lebensm. Unters. Forsch.*, 1993, **196**, 307–328.

82. C. Bicchi, V. Manzin, A. D'Amato and P. Rubiolo, *Flav. Fragr. J.*, 1995, **10**, 127–137.

83. C. Bicchi, A. D'Amato and P. Rubiolo, *J. Chromatogr. A*, 1999, **843**, 99–121.

84. W. A. König, C. Fricke, Y. Saritas, B. Momeni and G. Hohenfeld, *J. High Resolut. Chromatogr.*, 1997, **20**, 55–61.

85. P. Cozzini, P. Domiano, P. C. Musini, G. Palla and E. Zanardi, *J. Inclusion Phenom. Mol. Recognit. Chem.*, 1996, **26**, 295–302.

86. C. Bicchi, C. Balbo, A. D'Amato, V. Mazin, P. Schreier, A. Rozenblum and P. Brunerie, *J. High Resolut. Chromatogr.*, 1998, **21**, 103–106.

87. C. Bicchi, A. D'Amato, V. Mazin, A. Galli and M. Galli, *J. High Resolut. Chromatogr.*, 1997, **20**, 493–498.

88. C. Bicchi, G. Artuffo, A. D'Amato, V. Manzin, A. Galli and M. Galli, *J. High Resolut. Chromatogr.*, 1993, **16**, 209–214.

89. A. Dietrich, B. Maas and A. Mosandl, *J. High Resolut. Chromatogr.*, 1995, **18**, 152–156.

90. F. Kobor, K. Angermund and G. Schomburg, *J. High Resolut. Chromatogr.*, 1993, **16**, 299–311.

91. B. E. Kim, S. H. Lee, K. S. Park, K. P. Lee and J. H. Park, *J. High Resolut. Chromatogr.*, 1997, **20**, 208–212.

92. M. Jung and V. Schurig, *J. High Resolut. Chromatogr.*, 1993, **16**, 289–298.

93. J. S. Bradshaw, Z. Chen, G. Yi, B. E. Rossiter, A. Malik, D. Pyo, H. Yun, D. R. Black, S. S. Zimmermann, M. L. Lee, W. Tong and V. T. D'Souza, *Anal. Chem.*, 1995, **67**, 4437–4439.

94. L. Lindquist and P.-E. Jansson, *J. Chromatogr. A*, 1997, **767**, 325–329.

95. R. Hiltunen and I. Laakso, *Flav. Fragr. J.*, 1995, **10**, 203–210.

96. K. Sjödin, M. Persson, A.-K. Borg-Karlsson and T. Norin, *Phytochemistry*, 1996, **41**, 439–445.

97. M. Persson, K. Sjödin, A.-K. Borg-Karlsson, T. Norin and I. Ekberg, *Phytochemistry*, 1996, **42**, 1289–1297.

98. M. Pietsch and W. A. König, *J. High Resolut. Chromatogr.*, 1997, **20**, 257–260.

99. P. Hennig, A. Steinborn and W. Engewald, *Chromatographia*, 1994, **38**, 689–693.

100. H. Knapp, M. Straubinger, S. Fornari, N. Oka, N. Watanabe and P. Winterhalter, *J. Agric. Food Chem.*, 1998, **46**, 1966–1970.

101. L. Mondello, M. Catalfamo, A. R. Proteggente, I. Bonaccorsi and G. Dugo, *J. Agric. Food Chem.*, 1998, **46**, 54–61.

102. T. J. Betts, *J. Chromatogr. A*, 1994, **678**, 370–374.

103. W. A. König, A. Rieck, I. Hardt, B. Gehrcke, K.-H. Kubeczka and H. Muhle, *J. High Resolut. Chromatogr.*, 1994, **17**, 315–320.

104. I. H. Hardt, A. Rieck, C. Fricke and W. A. König, *Flav. Fragr. J.*, 1995, **10**, 165–171.

105. W. A. König, A. Rieck, Y. Saritas, I. H. Hardt and K.-H. Kubeczka, *Phytochemistry*, 1996, **42**, 461–464.

106. A. Rieck, N. Bülow and W. A. König, *Phytochemistry*, 1995, **40**, 847–851.

107. W. A. König, N. Bülow, C. Fricke, S. Melching, A. Rieck and H. Muhle, *Phytochemistry*, 1996, **43**, 629–633.

108. Y. Saritas, N. Bülow, C. Fricke, W. A. König and H. Muhle, *Phytochemistry*, 1998, **48**, 1019–1023.

109. A. Rieck and W. A. König, *Phytochemistry*, 1996, **43**, 1055–1056.

110. A. Rieck, N. Bülow, S. Jung, Y. Saritas and W. A. König, *Phytochemistry*, 1997, **44**, 453–457.

111. A. Rieck, N. Bülow, C. Fricke, Y. Saritas and W. A. König, *Phytochemistry*, 1997, **45**, 195–197.

112. C. Fricke, A. Rieck, I. H. Hardt, W. A. König and H. Muhle, *Phytochemistry*, 1995, **39**, 1119–1121.

113. T. Kajiwara, Y. Akakabe, K. Matsui, K. Kodama, H. Koga and T. Nagakura, *Phytochemistry*, 1997, **45**, 529–532.

114. J. C. R. Demyttenaere and H. M. Willemen, *Phytochemistry*, 1998, **47**, 1029–1036.

115. S. Melching, A. Blume, W. A. König and H. Muhle, *Phytochemistry*, 1998, **48**, 661–664.

116. S. Melching, N. Bülow, K. Wihstutz, S. Jung and W. A. König, *Phytochemistry*, 1997, **44**, 1291–1296.

117. A. O. Oyedeji, O. Ekundayo, M. M. Sonwa, C. Fricke and W. A. König, *Phytochemistry*, 1998, **48**, 657–660.

118. X. Wang, C. Jia and H. Wan, *J. Chromatogr. Sci.*, 1995, **33**, 22–25.

119. H. Casabianca, J. B. Graff, V. Faugier, F. Fleig and C. Grenier, *J. High Resolut. Chromatogr.*, 1998, **21**, 107–112.

120. C. Frank, A. Dietrich, U. Kremer and A. Mosandl, *J. Agric. Food Chem.*, 1995, **43**, 1634–1637.

121. M. Wüst, T. Beck and A. Mosandl, *J. Agric. Food Chem.*, 1998, **46**, 3225–3229.

122. S. Faulhaber, U. Hener and A. Mosandl, *J. Agric. Food Chem.*, 1997, **45**, 4719–4725.

123. D. Bartschat, C. Kuntzsch, M. Heil, A. Schittrigkeit, K. Schumacher, M. Mang, A. Mosandt and R. Kaiser, *Phytochem. Anal.*, 1997, **8**, 159–199.

124. H. Casabianca and J. B. Graff, *J. Chromatogr. A*, 1994, **684**, 360–365.

125. A. L. Morales, D. Albarracin, J. Rodriguez, C. Duque, L. E. Riano and J. Espitia, *J. High Resolut. Chromatogr.*, 1996, **19**, 585–587.

126. T. König, C. Ruff, M. Kleinschnitz, P. Schreier, N. Fischer and W. Neugebauer, *J. High Resolut. Chromatogr.*, 1998, **21**, 371–372.

127. B. Weber, A. Dietrich, B. Maas, A. Marx, J. Olk and A. Mosandl, *Z. Lebensm. Unters. Forsch.*, 1994, **199**, 48–50.

128. D. Krajewski, C. Duque and P. Schreier, *Phytochemistry*, 1997, **45**, 1627–1631.

129. A. Kaunzinger, M. Wüst, H. Grobmiller, S. Burow, U. Hemmrich, A. Dietrich, T. Beck, U. Hener, A. Mosandl and A. Rapp, *Z. Lebensm. Unters. Forsch.*, 1996, **203**, 499–500.

130. C. Dietrich, T. Beuerle, B. Withopf, P. Schreier, P. Brunerie, C. Bicchi and W. Schwab, *J. Agric. Food Chem.*, 1997, **45**, 3178–3182.

131. D. Häring, T. König, B. Withopf, M. Herderich and P. Brunerie, *J. High Resolut. Chromatogr.*, 1997, **20**, 135–354.

132. G. G. G. Manzardo, S. Kürsteiner-Laube and D. Perrin, *Z. Lebensm. Unters. Forsch.*, 1996, **203**, 501–506.

133. D. Bartschat, B. Maas, S. Smietana and A. Mosandl, *Phytochem. Anal.*, 1996, **7**, 131–135.

134. D. Bartschat, T. Beck and A. Mosandl, *J. Agric. Food Chem.*, 1997, **45**, 4554–4557.

135. D. Lehmann, B. Maas and A. Mosandl, *Z. Lebensm. Unters. Forsch.*, 1995, **201**, 55–61.

136. W. M. Coleman III, T. A. Perfetti and R. L. Suber, Jr. *J. Chromatogr. Sci.*, 1998, **36**, 318–321.

137. G. P. Blanch and G. J. Nicholson, *J. Chromatogr. Sci.*, 1998, **36**, 37–43.

138. H. J. Bouwmeester, J. A. R. Davies and H. Toxopeus, *J. Agric. Food Chem.*, 1995, **43**, 3057–3064.

139. S. Reichert, M. Wüst, T. Beck and A. Mosandl, *J. High Resolut. Chromatogr.*, 1998, **21**, 185–188.

140. W. Neugebauer and P. Schreier, *J. Agric. Food Chem.*, 1995, **43**, 1647–1653.

141. R. Reinhardt, A. Steinborn, W. Engewald, K. Anhalt and K. Schulze, *J. Chromatogr. A*, 1995, **695**, 475–484.

142. A. Steinborn, R. Reinhardt, W. Engewald, K. Wyssuwa and K. Schulze, *J. Chromatogr. A*, 1995, **697**, 485–494.

143. G. Fuchs and M. Perrut, *J. Chromatogr. A*, 1994, **658**, 437–443.

144. D. Bartschat and A. Mosandl, *Z. Lebensm. Unters. Forsch.*, 1996, **202**, 266–269.

145. G. Schmidt, W. Neugebauer, P. Winterhalter and P. Schreier, *J. Agric. Food Chem.*, 1995, **43**, 1898–1902.

146. B. Weber and A. Mosandl, *Z. Lebensm. Unters. Forsch.*, 1997, **204**, 194–197.

147. I. Roling, H.-G. Schmarr, W. Eisenreich and K.-H. Engel, *J. Agric. Food Chem.*, 1998, **46**, 668–672.

148. A. Mosandl, *Kontakte (Darmstadt)*, 1992, 38–49.

149. A.-K. Borg-Karlson, C. R. Unelius, I. Valterová and L. A. Nillson, *Phytochemistry*, 1996, **41**, 1477–1483.

150. B. Maas, A. Dietrich and A. Mosandl, *J. High Resolut. Chromatogr.*, 1994, **17**, 109–115.

151. B. Maas, A. Dietrich and A. Mosandl, *J. High Resolut. Chromatogr.*, 1994, **17**, 169–173.

152. R. Reinhardt, M. Richter, P. P. Mager, P. Hennig and W. Engewald, *Chromatographia*, 1996, **43**, 187–194.

153. D. Laudert, P. Hennig, B. A. Stelmach, A. Müller, K. Andert and E. W. Weiler, *Anal. Biochem.*, 1997, **246**, 211–217.

154. J. Krupcik, E. Benická, P. Májek, I. Skacáni and P. Sandra, *J. Chromatogr. A*, 1994, **665**, 175–184.

155. T. Takeichi, H. Toriyama, S. Shimura, Y. Takayama and M. Morikawa, *J. High Resolut. Chromatogr.*, 1995, **18**, 179–189.

156. M. Jung, M. Fluck and V. Schurig, *Chirality*, 1994, **6**, 510–512.

157. B. H. Hoff and T. Anthonsen, *Chirality*, 1999, **11**, 760–767.

158. U. Meierhenrich, W. H.–P. Thiemann and H. Rosenbauer, *Chirality*, 1999, **11**, 575–582.

159. T. Beier and H.-D. Höltje, *J. Chromatogr. B*, 1998, **708**, 1–20.

160. W. M. Buda, K. Jacques, A. Venema and P. Sandra, *Fresenius' J. Anal. Chem.*, 1995, **352**, 679–683.

161. A. Berthod, E. Y. Zhou, K. Le and D. W. Armstrong, *Anal. Chem.*, 1995, **67**, 849–857.

162. R. Reinhardt, W. Engewald, U. Himmelreich, B. Christian and B. Koppen-hoefer, *J. Chromatogr. Sci.*, 1995, **33**, 236–243.

163. K. Jacques, W. M. Buda, A. Venema and P. Sandra, *J. Chromatogr. A*, 1994, **666**, 131–136.

164. F. Kobor and G. Schomburg, *J. High Resolut. Chromatogr.*, 1993, **16**, 693–699.

165. D. U. Staerk, A. Shitangkoon and G. Vigh, *J. Chromatogr. A*, 1995, **702**, 251–257.

166. W. M. Buda, K. Jacques, A. Vanema and P. Sandra, *Fres. J. Anal. Che.*, 1995, **352**, 679–683.

167. A. Berthold, E. Y. Zhou, K. Le and D. W. Armstrong, *Anal. Chem.*, 1995, **67**, 849–857.

168. R. Reinhardt, W. Engewald, U. Himmelreich, B. Christian and B. Koppen-hoefer, *J. Chromatogr. Sci.*, 1995, **33**, 236–243.

169. K. Jacques, W. M. Buda, A. Venema and P. Sandra, *J. Chromatogr. A*, 1994, **666**, 131–136.

170. F. Kobor and G. Schomburg, *J. High Resol. Chematogr.*, 1993, **16**, 693–699.

171. D. U. Staerk, A. Shitangkoon and G. Vigh, *J. Chromatogr. A*, 1995, **702**, 251–257.

Use of Cyclodextrins in Supercritical Fluid Chromatography

Because of the considerable advantages (extended molecular weight range compared to GC, separation of thermally labile compounds at lower temperatures, sensitive detection of compounds without chromophores, use of open-tubular and packed columns, simple eluent composition, rapid column equilibration) super-critical fluid chromatography (SFC) combined with CDs has found application in the analysis of bioactive compounds and other organic molecules particularly of chiral analytes.

CD–oligosiloxane copolymers were frequently used for enantiomeric separation in open tubular column SFC.[1] Separations were performed with a density program from 0.30 g mL^{-1} at a rate of 0.010 g mL^{-1} after 1 min of isopycnic period (Figure 3.1). It was stated that CD-oligosiloxane copolymers can be successfully used for the enantiomeric separation of a wide variety of racemic compounds in open tubular column SFC.

Naphthylethylcarbamoylated-β-CD stationary phases (NEC-CD CSP) have also been prepared and their enantioselectivity tested in liquid, sub- and super-critical fluid chromatography.[2] Separations were performed on (R)- and (S)-naphthylethylcarbamoylated-β-CD columns by normal and reversed-phase HPLC using respectively, hexane-2-propanol and 1% triethylammonium acetate, and SFC. The enantiomers separated well with both HPLC and SFC, but the analysis time of the SFC method was shorter (Figure 3.2). It was stated that the high versatility of SFC makes it a valid alternative to both normal and reversed-phase HPLC for the enantiomeric separation of racemic compounds.

Permethyl-substituted β-CD polysiloxane stationary phases have also been synthesised and showed excellent enantioselectivity for a wide variety of racemic compounds on fused silica capillary columns under GC and SFC conditions[3] (Figure 3.3). It was stated that the permethyl-substituted β-CD polysiloxane stationary phases are suitable.

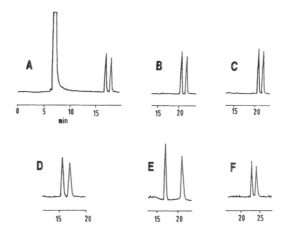

Figure 3.1 *SFC separation of the enantiomers of diethyl tartrate (A), 2-phenylcyclohexanol (B), 1-phenylethanol (C), ibuprofen (D), pantolactone (E), and 1,2-diphenyl-1,2-ethanediol (F)*
(Reprinted with permission from ref. 1)

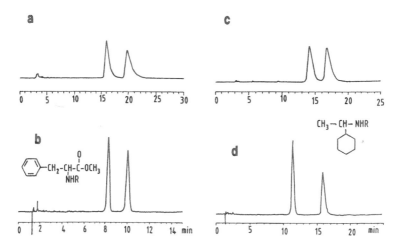

Figure 3.2 *Separation of N-(3,5-dinitrobenzoyl)-D,L-phenylalanine (a, b) and N-(3,5-dinitrobenzoyl)-R,S-1-cyclohexylethylamine (c, d) on the S-NEC-CD CSP. Chromatographic conditions: (a) hexane–2-propanol (60:40), 1.0 mL min⁻¹, (b) carbon dioxide–methanol (85:15), 2.0 mL min⁻¹, 30 °C, 15 MPa, (c) hexane–2-propanol (80:20), 1.0 mL min⁻¹, (b) carbon dioxide–methanol (90:10), 2.0 mL min⁻¹, 30 °C, 15 MPa. Column: Cyclobond I 2000 RN (25 × 0.46 mm, 5 μm particle size)*
(Reprinted with permission from ref. 2)

Octakis-(3-*O*-butanoyl-2,6-di-*O*-*n*-pentyl)-γ-cyclodextrin (3-butanoyl-2,6-pentyl-γ-CD) was covalently bonded to a polysiloxane using an octamethyl spacer and the enantioselectivity of this stationary phase (Chirasil-γ-Dex) was tested under GC and SFC conditions and showed good enantiomeric selectivity (Figure

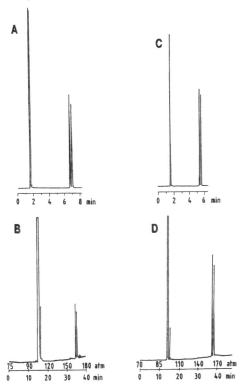

Figure 3.3 *Separation of (±)-trans-1,2-cyclohexanediol (A, B) and (±)-α-trifluoromethyl)-benzyl alcohol (C, D) enantiomers on β-cyclodextrin-bound polysiloxane stationary phase by (A, C) GC and (B, D) SFC. Conditions: (A, C) 30 m × 250 µm ID column, 0.25 µm film thickness; 140 °C; helium carrier gas, FID; (B) 15 m × 50 µm ID column, 0.20 µm film thickness; 60 °C; pressure program from 75 atm, 5 min hold, at a rate of 3 atm min⁻¹, to 180 atm; (D) 15 m × 50 µm ID column, 0.20 µm film thickness; 60 °C; pressure program from 75 atm, 5 min hold, at a rate of 3 atm min⁻¹, to 200 atm*
(Reprinted with permission from ref. 3)

3.4).[4] Because of the excellent enantioselectivity the chiral stationary phase Chirasil-γ-Dex was proposed for the separation of racemic compounds in GC and SFC.

Permethylated allyl-substituted β-CD has been covalently bonded to an organosiloxane polymer and the mixture has been immobilised in fused silica capillaries. The capillaries have been employed for GC, CE and SFC determination of racemic model compounds. SFC chiral analysis of 1-aminoindan was carried out at 60 °C using CO_2 without additives. Initial density was 0.35 g ml⁻¹; after 1 min a density program (0.01 g ml⁻¹ min⁻¹) was initiated.[5] Another synthetic method has been applied for the immobilisation of octakis-(2,6-di-O-methyl-3-O-pentyl)-γ-CD to narrow-bore fused silica capillaries and the chiral separation capacity of capillaries has been tested in GC and SFC. SFC successfully separated

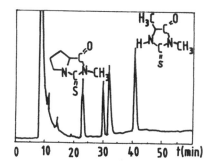
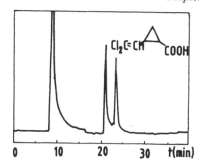

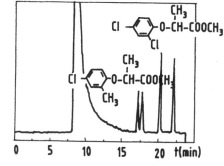
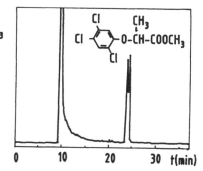

Figure 3.4 *SFC separation of MTH-proline and MTH-alanine (top left), cis-permethrinic acid (top right), mecopropmethylester and dichlorpropmethylester (bottom left) and fenopropmethylester (bottom right). The fused-silica capillary column (5 m × 50 μm ID) was coated with immobilized Chirasil-γ-Dex (40%, w/w with stepwise gradient elution methanol–dichloromethane*
(Reprinted with permission from ref. 4)

the enantiomers of chlorthenoxazine (80 °C, 160 bar CO_2), *o,p'*-DDT (70 °C, 140 bar CO_2), 2-ethylsuccinic acid (80 °C, 160 bar CO_2) and phenylhydantoin (80 °C).[6] Enantiomers of ten benzodiazepine racemates have been separated by GC, open tubular SFC and subcritical fluid chromatography using permethylated β-CD as chiral selector. Good enantiomer separation of dihydrodiazepam, oxazepam, lorazepam and ethyl loflazepate has been achieved by SFC.[7]

References

1. G. Yi, J. S. Bradshaw, B. E. Rossiter, S. L. Reese, P. Petersson, K. E. Markides and M. L. Lee, *J. Org. Chem.*, 1993, **58**, 2561–2565.
2. K. L. Williams, L. C. Sander and S. A. Wise,. *J. Chromatogr. A*, 1996, **746**, 91–101.

3. G. Yi, J. S. Bradshaw, B. E. Rossiter, A. Malik, W. Li and M. L. Lee, *J. Org. Chem.*, 1993, **58**, 4844–4850.
4. H. Grosenick and V. Schurig, *J. Chromatogr. A*, 1997, **761**, 181–193.
5. D. W. Armstrong, Y. Tang, T. Ward and M. Nichols, *Anal. Chem.*, 1993, **65**, 1114–1117.
6. J. Dönnecke, W. A. König, O. Gyllenhaal, J. Cessmad and C. Schulze, *J. High Resolut. Chromatogr.*, 1994, **17**, 779–783.
7. S. R. Almquist, P. Petersson, W. Walther and K, E. Markides, J. Chromatogr. A, 1994, **679**, 139–146.

CHAPTER 4

Use of Cyclodextrins in Liquid Chromatography

Cyclodextrins and cyclodextrin derivatives have found extensive application in liquid chromatography (LC). As in GC they have been employed to improve the separation of compounds with highly similar chemical structures (mainly positional and optical isomers). However, there is a difference in principle between the use of CDs in GC and in liquid chromatographic techniques. While CDs can be used only as stationary phases in GC, in LC they can be employed not only as modifiers of the stationary phase but also as additives in the mobile phase.

1 Thin-layer Chromatography

The separation capacity of TLC, even of HPTLC, is generally lower than that of similar HPLC methods. This has considerably limited the application of CDs in TLC for the enantiomeric separation of β-CD mixtures. However, the rapidity, the possibility of parallel analyses and the low instrumental costs make TLC advantageous for the determination of the strength of interactions between CDs, CD derivatives and guest molecules.

Modification of Retention Behaviour

Two-dimensional HPTLC has been employed for the separation of unconjugated and conjugated bile acids on C_{18}-bonded silica PTLC plates.[1] The plates were developed in the first direction with methanol−water mobile phase and then in the second direction with methanol−water containing 5 mM heptakis-(2,6-di-*O*-methyl)-β-CD (DIMEB). The method was developed with authentic standard compounds and was employed for the separation of glycine-conjugated bile acids in a human bileon. The results showed that the retention of solutes decreased in the presence of DIMEB. The differences in retention among the solutes made possible the use of DIMEB for the improvement of the separation capacity of the HPTLC system. The effect of the addition of α-, β-, γ- and hydroxypropyl β-CDs to the mobile phase on the reversed-phase mobility of nitroanilines and their derivatives has been studied in detail. It has been established that retention of solutes decreased in both silica and polyamide layers.[2]

Determination of the Strength of Cyclodextrin–Solute Interaction

The interaction of α-, β- and γ-CDs with *n*-alcohols has been studied by thermostated RP-TLC and the enthalpy and entropy of the interaction was

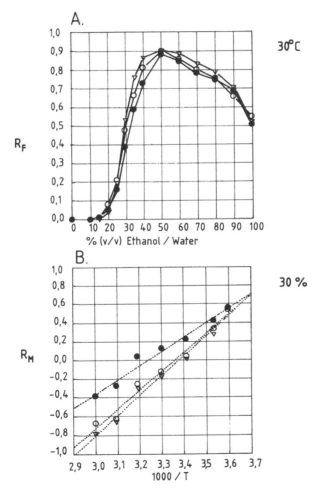

Figure 4.1 *Relationship between R_F (R_M) values of α-CD (○), β-CD (●) and γ-CD (▽) versus concentration of ethanol in water (graph A) and reciprocal of absolute temperature using 30% ethanol–water mixture as a mobile phase (graph B)* (Reprinted with permission from ref. 3)

calculated from the temperature dependence of the retention.[3] CDs showed anomalous retention behaviour. Their retention decreased at the lower concentration range of alcohols, reached a minimum, then increased again at higher concentrations of alcohol in the mobile phase (Figure 4.1). This anomaly was tentatively explained by the supposition that CDs form inclusion complexes with the *n*-alcohols and that the retention behaviour of the complex deviates from that of uncomplexed CDs. Not only the temperature dependence of the retention but also the dependence of retention on the concentration of CD in the eluent was employed for the determination of the strength of the host–guest interaction. The relationship between the R_M values, calculated from R_F by $R_M = \log(1/R_F - 1)$, and the concentrations of organic modifier and CDs was described by

$$R_M = R_{M0} + b_1 C_1 + b_2 C_2 \qquad (4.1)$$

where $R_M = R_M$ value for a solute determined at given concentrations of organic modifier and CD; $R_{M0} = R_M$ value extrapolated to zero concentrations of organic modifier and CD; $b_1 =$ decrease in the R_M value caused by a 1% increase in the concentration of the organic modifier in the eluent; $b_2 =$ decrease in the R_M value caused by a $1\,\mathrm{mg\,mL^{-1}}$ concentration change of CD in the eluent (related to the relative strength of interaction); C_1 and $C_2 =$ concentrations of organic modifier and CD, respectively. It was established that the method was suitable for the determination of the relative strength of interaction, the b_2 values being linearly correlated with the stability constant determined by other methods. A wide variety of host–guest pairs were studied with the RPTLC procedure outlined above: anticancer drugs with α-[4] and γ-CD,[5] with carboxymethyl-β-CD,[6] commercial pesticides HP-β-CD,[7] antisense nucleotides HP-β-CD,[8] polychlorinated biphenyls γ-CD,[9] steroidal drugs with carboxymethyl-γ-CD,[10] γ-CD,[11] HP-β-CD,[12] carboxymethyl-β-CD[13] and a water-soluble β-CD polymer,[14] anionic surfactants HP-β-CD[15] and antiviral nucleosides γ-CD.[16] It was established that in the majority of cases both the dimensions of the CD cavity and the type of substituents on the outer sphere of the CD exert a considerable impact on the strength of the host–guest interaction.[17]

Separation of positional and Optical Isomers

TLC methods suitable for chiral discrimination including the use of CDs as eluent additives have been reviewed.[18] The separation of the enantiomers of fluoxetine (FL), an antidepressant drug, its metabolites, norfluoxetinee (NL) and promethazine (PR) and isopromethazine (i-PR) was recently reported.[19] Separations were performed on silica TLC plates impregnated with 2.5% paraffin oil with a mobile phase consisting of a mixture of methanol–aqueous triethylamine buffer, pH being adjusted with glacial acetic acid. In order to enhance the separation power of the system, β-CD and hydroxypropyl-β-CD (HP-β-CD) were added separately to the mobile phase. It was established that both optical and positional isomers can be successfully separated by RP-TLC with HP-β-CD as eluent additive.

Enantiomers of aromatic amino acids and aromatic amino alcohols have also been separated by β-CDs as a chiral selector.[20] It was found that DL-3,4-dihydroxyphenylalanine, DL-*p*-hydroxyphenylglycine, DL-thyronine, DL-epinephrine and DL-isopropylepinephrine can also be separated by the mobile phase methanol–formic acid–0.2 M α-CD solution of urea (7:1:2).

2 High Performance Liquid Chromatography

Because of their versatility, sensitivity and reproducibility HPLC methods using CD additives in both stationary and mobile phases have been extensively used for the separation of optical and positional isomers. The theory and practical applications have been reviewed.[21,22]

Cyclodextrin Stationary Phases

Determination of the Strength of interaction Between CD-Bonded Stationary Phase and solutes

The elucidation of the influence of various physicochemical parameters of both host and guest molecules may help the better understanding of the character of interaction between stationary phases and solutes resulting in rational experimental design and more rapid optimisation of any practical separation problems. However, the number of studies dealing with the assessment of these relationships between the characteristics of the stationary phase and the chemical structure of the solute is surprisingly low.

Six imidazole derivatives were employed as model solutes for the study of the retention mechanism of a β-CD-bonded stationary phase[23] with mobile phases consisting of 2-propanol-buffer mixtures and water. Differential scanning calorimetry (DSC) and thermogravimetric analysis (TGA) were used for the study of the interaction between the β-CD-bonded stationary phase, silica, β-CD and mobile phases in the temperature range of 0–250 °C. Thermodynamic parameters of transfer of solute from the mobile phase to the stationary phase were calculated by

$$\ln k' = H°/RT + S°^* \tag{4.2}$$

$$S°^* = S°/R + \ln\varphi \tag{4.3}$$

where $-H°$ and $S°$ are the enthalpy and entropy of transfer, T is the temperature, R is the gas constant and φ is the phase ratio of the column (volume of stationary phase divided by the volume of the mobile phase). It was found that the relationship between the ln k values and the concentration of water in the eluent is not linear, but can be described with a second-order polynomial. The correlation between ln k and $1/T$ was linear at pH 7.0 and 7.5 but showed marked non-linearity at pH 6.5, 8.0 and 8.5.

The non-linearity of the relationship ln k' *versus* $1/T$ at pH 6.5, 8.0 and 8.5 was explained by the supposition that the hydrophobic character of the cavity of β-CD is pH dependent resulting in different binding capacity. DSC and TGA measurements of the β-CD-bonded stationary phase found a phase transition at 43 °C at the same pH. The binding constants of methyl benzoate anions with β-CD were determined by both CE and RP-HPLC using a β-CD bonded column.[24] It was established that the binding constants determined by both methods are similar; therefore they can be successfully used for the measurement of the stability of inclusion complexes between solutes and β-CD.

Separation of Positional and Optical Isomers of Pesticides and Other Environmental Pollutants

Chiral phenoxypropionate herbicides have been separated by both HPLC and GC using heptakis-(2,3,6-tri-*O*-methyl)-β-CD (TRIMEB),[25] such as fluazifop and diclofop esters (Figure 4.2). However, it was established that isocratic elution cannot be used for the determination of these pesticides in soil, as coextracted impurities coeluted with the solutes. GC separations were performed on a fused silica capillary column coated with TRIMEB dissolved in OV-1701, and a comparison of HPLC and GC data indicated the higher separation power of HPLC method.

A silica column with permethylated β-CD covalently bonded to the stationary

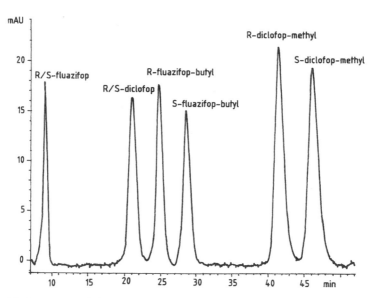

Figure 4.2 *Separation of fluazifop, diclofop and the enantiomers of corresponding butyl and methyl esters using chiral permethylated β-cyclodextrin HPLC column (eluent: 70% methanol buffered at pH 4 with 0.4% triethyl ammonium acetate, flow-rate: 0.4 mL min⁻¹, injection volume: 20 µL, detection: 233 nm)*
(Reprinted with permission from ref. 25)

phase has been employed for the separation of polychlorinated biphenyl (PCB) atropisomers (Figure 4.3)[26] but the method was not able to separate every atropisomer pair. Another HPLC method using permethylated-β-CD-derivatised silica stationary phase and subambient temperature (0 °C) has also been suitable for the enantiomer separation of PCB atropisomers.[27] The influence of common chromatographic conditions on the retention of environmental pollutants but also that of pressure has been studied in detail.[28] The study was motivated by the observation that the pressure modifies the ionisation state of solute and its distribution between the stationary and mobile phases. As the outer sphere of the CD cavity is hydrophilic the change of ionisation may influence the retention on CD-bonded stationary phase. Column pressure exerted a considerable impact on the separation of nitrophenol isomers (Figure 4.4). The data suggested that the host−guest complexes between the stationary phase and solutes were more dissociated under pressure, with the effect being higher for more ionised solutes.

The retention behaviour of positional isomers was determined on two different β-CD stationary phases and the results were compared.[29] Native β-CD and heptakis-2,3-*O*-dimethyl-β-CD-bonded silica stationary phases and buffered and

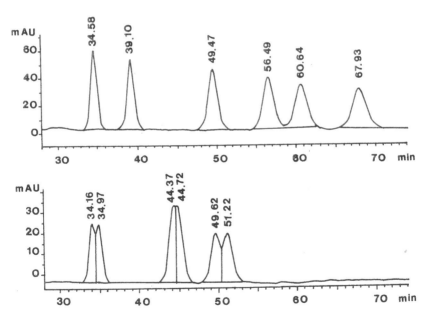

Figure 4.3 *UV chromatograms of fully (top) and partially (bottom) separated atropisomers of PCB. The fully resolved peaks originate from PCB84 (35 and 39 min), PCB174 (49 and 56 min), and PCB 196 (61 and 68 min). The partially resolved peaks are from PCB91 (ca. 34 min), PCB95 (ca. 44 min), and PCB171 (ca. 50 min). Column: 5 μm Nucleodex β PM and 5 μm Nucleosit 100 with bonded chiral selectors*
(Reprinted with permission from ref. 26)

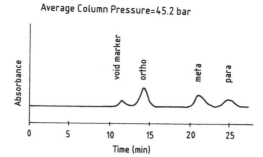

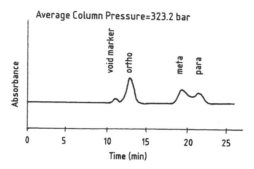

Figure 4.4 *Effect of pressure on the separation of a mixture of nitrophenol isomers on a capillary column packed with β-cyclodextrin bonded phase and eluted with 20:80 (v/v) methanol–10 mM Tris buffer mobile phase at pH 7.50. Column flow-rate 1.5 μL min⁻¹; T = 24 °C*
(Reprinted with permission from ref. 28)

unbuffered methanol–water mobile phases were employed. The data show that neutral compounds, substituted aniline and phenol derivatives are retained more strongly on the methylated β-CD phase while the retention of substituted carboxylic acids was longer on the native β-CD column. Interestingly, the capacity of both columns for the separation of positional isomers was similar.

As well as chiral stationary phases with covalently bonded CDs on the surface, polymer-coated silicas have also been prepared with β-CD containing poly(vinylamine) and the pore size distributions, surface areas and polymer contents of the new stationary phases were determined.[30] The capacity of the new phases for the separation of positional isomers was studied in detail.[31] It was found that the retention capacity of the stationary phases increased with increasing amount of polymer on the silica surface and this was accompanied by peak broadening. The stationary phases were proposed for the separation of nitrophenol, nitrobenzoic acid and cresol isomers. Similar CD-bonded stationary phases have been employed for the separation of structural isomers of disubstituted benzene derivatives.[32] HPLC using a permethylated β-CD column has been employed for the enantiomeric separation of (±)-*cis*-chlordane, (±)-*trans*-chlordane, (±)-chlor-

Immobilsed β-CD has also been employed for the enantiomeric separation of amphetamine, methamphetamine and ring substituted amphetamines.[42]

A β-CD-bonded stationary phase has been employed for the enantiomeric separation of the antidepresant drug fluoxetine (FL) and its metabolite norfluoxetine (NR).[43] The enantioselectivity decreased linearly with increasing ionic strength of the mobile phase, which was tentatively explained by the supposition that the TEA molecules compete with the drug molecules for the cavities of the β-CD. The increase of pH also increased the selectivity and resolution of enantiomers. It was assumed that this effect may be due to ion pair formation between the polar substructures of solutes and the buffer ions. The type of organic modifier also influenced the separation capacity of the HPLC system which was attributed to the formation of ternary complexes consisting of β-CD-solute-organic modifier. The best chromatographic conditions for the separation of FL and NR enantiomers were: MeOH–buffer, 67/33 (v/v), 0.5 TEAA (w/v); pH 6.5; flow-rate 0.8 mL min^{-1}.

The efficiencies of various chromatographic techniques, capillary electrophoresis, HPLC with chiral mobile phase additives and with a chiral stationary phases, were compared for the separation of FL and NR enantiomers.[44] HPLC measurements were performed on a phenyl and a Cyclobond I column using ACN–water–TEA mobile phases and fluorescence detection. CE separations were carried out on a fused silica capillary applying 20 kV voltage and UV detection at 230 nm. CDs were added to the mobile phases in the LC and CE methods and the stability constants were calculated for the dependence of retention on the concentration. The chromatograms (Figure 4.6) show that each method can be used for the enantiomeric separation of FL; however, the efficiency of the chiral separations may be different using the same buffer. As the binding processes and binding constants were similar in both HPLC–MPA and CE the data indicated that the type of CD or CD derivative exerted a considerable influence on the efficiency of the chiral separation.

The effect of column pressure on the enantiomeric separation of various chiral pharmaceuticals has been studied in detail.[45] The data suggested that an increase in plate height at higher pressures accounts for the decrease of chiral selectivity. It was emphasised that the appropriate control of column pressure may enhance the efficacy of enantiomeric separations.

(S)-and (R)-propanolol in human plasma and urine have been separated and quantitatively determined using a Cyclobond I column in a pharmacokinetic study of β-CD propanolol.[46] Not only reversed-phase conditions but also polar-organic mobile phase system have been used for the chiral separation of propanolol and other chiral pharmaceuticals.[47] A β-CD-bonded stationary phase was employed for the enantioselective separation of myoinositol phosphates.[48]

The enantiomers of the local anaesthetic carbisocaine have also been separated on a β-CD stationary phase (Figure 4.7).[49] The enantiomeric separation of barbiturates and thiobarbiturates has been performed on β-CD, hydroxypropyl-β-CD and β-CD polymer coated columns.[50] Stationary phases containing native CDs and various CD derivatives have been used for the separation of the enantiomers of a new prostacyclin analogue, beraprost sodium. It was found that

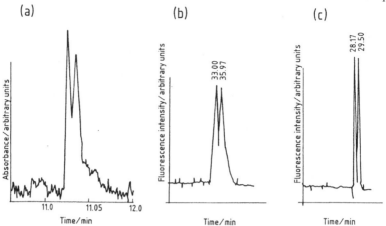

Figure 4.6 *Separations of fluoxetine enantiomers using (a) CE with 1 mM β-CD added to buffer, (b) LC with 1 mM β-CD added to mobile phase and (c) LC with a Cyclobond I CSP. Buffer conditions 1% triethylammonium acetate, pH 5.5, 10% acetonitrile flow rate/detection*
(Reprinted with permission from ref. 44)

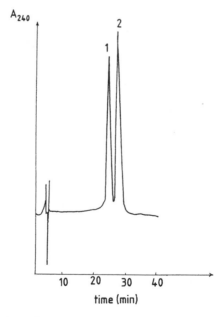

Figure 4.7 *The separation of racemic mixture of carbisocaine (0.1 mg mL⁻¹). Conditions: mobile phase 5% acetonitrile in water and 0.03% triethylamine, pH = 6.1; flow-rate 0.5 mg mL⁻¹; detection at 240 mm, volume of injection 10 μL. 1 = R(−) enantiomer of carbisocaine, 2 = S(+) enantiomer of carbisocaine*
(Reprinted with permission from ref. 49)

the best separation of the four isomers can be achieved on acetylated β-CD support.[51] The enantioseparation of five dihydropyridine calcium antagonists (amlodipine, nitredipine, isradipine, nimodipine, nisoldipine) has been obtained on a β-CD column and the energy of interaction was calculated.[52] Chiral selectivities of various CD-bonded silica columns have been compared in the analysis of non-steroidal anti-inflammatory drugs (fenoprofen, flurbiprofen, ibuprofen, ketoprofen, naproxen). Considerable differences were observed in the chiral selectivities of stationary phases.[53] Sulfated β-CD has also been applied for the enantiomeric separation of pharmaceuticals (antihistamines, antidepressants, *etc.*) and another chiral compounds and the sterical correspondence and electro-static interactions were found to govern the enantiomer separation.[54]

New chiral stationary phases based on β-CD using different spacers have been compared for the separation of antidepressants (oxazepam, lorazepam, *etc.*) and other chiral compounds as model solutes.[55] A comparison of the separation of the enantiomers of trimetoquinol, denopamine and timepidium by HPLC on β-CD-bonded stationary phase and by CE using CDs as electrolyte additives have found that the latter method has a higher separation capacity.[56]

Simulated moving-bed chromatography has been used for the preparative-scale separation of the enantiomers of precursors of a novel Ca-sensitising drug on various chiral stationary phases. It was found that the method is suitable for the optimisation of the chiral separation of enantiomer pairs.[57] Bare silica stationary phases and aqueous mobile phase containing native β-CD has been suitable for the enantiomeric separation of a β-CD serotonin receptor {($\pm$)-3-methyl-amino-methyl-3,4,5,6-tetrahydro-6-oxo-1*H*-azepino[5,4,3-*cd*]indole hydrochloride and its *N*-desmethyl derivative.[58] The separation of the enantiomers of 7-desmethyl-ormeloxifene by counter-current chromatography using sulfated β-CD as chiral selector has also been reported.[59]

Separation of Positional and Optical Isomers of Amino Acids and Related Compounds

Because of the biological importance of the discrimination between D- and L-amino acids much effort has been devoted to the development of HPLC methods suitable for the enantioselective separation of amino acids. Many derivatisation processes have been developed and successfully applied to facilitate separation and detection of amino acids. Thus, derivatisation with 9-fluorenylmethyl chloroformate (FMOC-Cl), FMOC-glycyl-Cl and FMOC-β-alanyl-Cl has been reported.[60] Enantiomeric separation of derivatised amino acids was performed on β-CD- and a γ-CD-bonded stationary phases using mobile phases consisting of ACN, TEA, MeOH, acetic acid and water. It was concluded from the data that the inclusion of glycyl and β-alanyl groups in the derivatives of amino acids increased the enantioselectivity of chiral separation. Retention was influenced by the type of solute, chiral selector and mobile phase additives.

Phenylcarbamoylated β-CDs, prepared by reacting β-CD with phenyl isocya-nate in different ratios, were packed into columns and the enantioselectivities were compared using phenylcarbamoylated amino acids (PTC-AAs) as solutes.[61] As the separation of PTC-AAs was insufficient, the solutes were pre-separated on a chiral RP-HPLC column. It was established that the bulkiness and hydrophobicity of the amino acid side chain increased the retention, and the separation factor also depended on the concentration of ammonium acetate in the mobile phase.

A 3-O-methyl-β-CD-bonded stationary phase and a β-CD column were compared for the enantiomeric separation of 2,4-dinitrophenyl (DNP) amino acids.[62]

The former column gave higher retentions and better enantioselectivity which was explained by the supposition that the cavity was more hydrophobic due to the presence of methyl group and forms more stable complexes with the hydrophobic solutes.

Application of selectively methylated γ-CD stationary phases for the enantiomeric separation of dansylamino acids was also reported.[63] Native γ-CD, octakis-(2-O-methyl)-γ-CD (2MM-γ-CD), octakis-(3-O-methyl)-γ-CD (3MM-γ-CD), and octakis-(2,3-di-O-methyl)-γ-CD (2,3DM-γ-CD) were covalently bonded to silica surface with an allyloxyethyl spacer arm. The separation depended markedly on the composition of the mobile phase and the type of solute to be separated. Resolution increased with decreasing concentration of methanol in the mobile phase. Native γ-CD and 3MM-γ-CD stationary phases showed good enantioselectivity while no resolution was observed on 2MM-γ-CD and 2,3DM-γ-CD. This result suggested that the secondary hydroxyl groups at 2-position is essential for chiral recognition. Mobile-phase pH also influenced the resolution. CD polymer-coated silica stationary phases have also been prepared and their enantioselectivity has been tested by using derivatised amino acids, barbiturates, and coumarins as model compounds.[64] EP-β-CDN$^+$ polymer was prepared by reacting native β-CD with epichlorohydrin and treating the reaction product with 2,3-epoxypropyltrimethylammonium chloride. Methylated and acetylated polymers were prepared from EP-β-CDN$^+$ and were adsorbed on silica support. Derivatisation of the β-CD polymer generally resulted in decreased enantioselectivity. This finding was tentatively attributed to the steric hindrances caused by the substituents.

The enantioselectivity of β-CD and heptakis-(2,3-O-dimethyl)-β-CD stationary phases was compared using DNS amino acids and other bioactive molecules as model compounds.[65] The majority of solutes were separated on methylated β-CD; however, in some cases the native β-CD-bonded stationary phase exhibited higher chiral separation capacity. The differences were explained by the assumption that the impact of steric interactions is more important on methylated β-CD while retention is mainly governed by hydrogen bonding on native β-CD. The character of the host–guest interaction accounted for the differences in the separation.

A chiral stationary phase with chemically bonded β-CD has been used for the analysis of dansylsulfonamide, β-naphthamide of β-naphthyl ether derivatives of amino acids and other chiral solutes. The preponderant role of the formation of inclusion complexes between the stationary phase and solutes has been emphasized.[66] Monoallylethoxylated β- and γ-CDs have been covalently bonded to a hydride modified or a 3-(2-tosyloxyethoxy)propyl silica surface, and the new stationary phases were tested using dansylamino acids as solutes. High differences have been observed between the enantioselectivity of the stationary phases.[67]

The effect of the derivatisation agent on the chiral separation of amino acids on a β-CD stationary phase has been studied in detail. Phenyl isothiocyanate, 5-dimethylaminonaphthalene-1-sulfonyl chloride, 4-dimethylaminoazobenzene-4′-sulfonyl chloride, 1-naphthyl isothiocyanate, 4-dimethylaminoazobenzene-4′-isothiocyanate, 4-dimethylamino-1-naphthyl isothiocyanate, and 6-aminoquinolyl-N-hydroxysuccinimidyl carbamate (AQC) were included in the experiments. The enantiomeric separation markedly depended on the type of amino acid and that of the derivatising reagent.[68] Another study used AQC as derivatising reagent and achieved the enantiomeric separation of 31 amino acids on various CD stationary phases (α-, β-, γ-, R,S-2-hydroxypropyl, acetylated β-, S-naphthylethyl-carba-mated, R-naphthylethyl-carbamated CD chiral selectors). The separation always depended on the solute–stationary phase interaction.[69] The same derivatising reagent and the same columns have been successfully used for the enantiomeric separation of di- and tri-peptides.[70] Glycyl di- and tri-peptides derivatised with 9-fluorenylmethyl chloroformate have also been separated on native CD-bonded phases.[71]

Separation of Positional and Optical Isomers of Miscellaneous Organic Compounds

The enantioselectivity of acetyl- and native-β-CD-bonded stationary phases was compared in the chiral separation of ten-vertex carboranes.[72] It was established that acetyl-β-CD-bonded stationary phase showed a better enantioselectivity than the native-β-CD-bonded phase (Figure 4.8). The differences in the chiral resolution were attributed to different positions of CH groups and the acidity of the molecules.

The separation of phosphorylated carbohydrates has also been carried out on a β-CD-bonded stationary phase.[73] Solutes were detected and identified with electrospray ionisation tandem mass spectrometry. It was assumed that the separation of phosphorylated carbohydrates was due to the anion exchange capacity of β-CD and not the formation of inclusion complexes between solutes and stationary phase. It was further stated that the method was suitable for the separation and identification of phosphorylated carbohydrates that are important in biosynthetic pathways.

Phenylcarbamate substituted β-CD derivatives have been also employed in normal-phase separation of a number of solutes (Figure 4.9).[74] Separations were carried out with a mobile phase containing hexane and 2-propanol in 98:2, 90:10 and 70:30 ratios v/v, and the separation improved with increasing degree of substitution. It was postulated that hydrogen bonding, π–π interactions and dipole stacking equally contribute to the efficiency of enantiomer separation.

Normal-phase chiral separation of α-, β-, γ- and δ-tocopherols and 5,7-dimethyltocol was performed on β- and γ-CD stationary phases and the marked effect of the composition of the mobile phase was observed.[75] Both saturated and unsaturated oligogalacturonic acids were separated on a β-CD bonded column using gradient elution. It was established that this stationary phase behaves as a

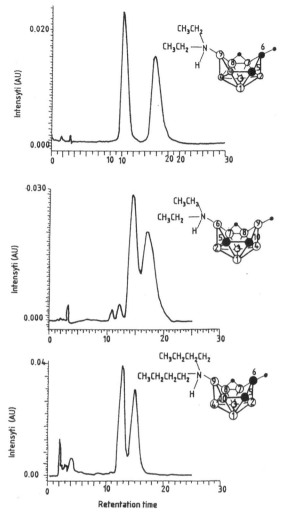

Figure 4.8 *Influence of the isomer structure and the size of alkyls on the secondary amino-group as ligand L in arachno-9-L-5,6-C₂B₈H₁₂ series on separation on bonded-acetyl-β-cyclodextrin column (125 × 4 mm ID); mobile phase: 55% methanol, flow-rate: 0.4 mg mL⁻¹; detection: UV DAD, fixed wavelength 254 nm* (Reprinted with permission from ref. 72)

traditional anion exchanger.[76] Anomeric saccharides have also been successfully separated on a β-CD-bonded stationary phase using ethyl acetate:methanol:water as mobile phase.[77] Enantiomers of salsolinol (1-methyl-6,7-dihydroxy-1,2,3,4-tetrahydroisoquinoline) and N-methylsalsolinol have been separated on CD-modified stationary phases. Because of the high sensitivity the method has been applied for the analysis of these alkaloids in human brain.[78] The positional

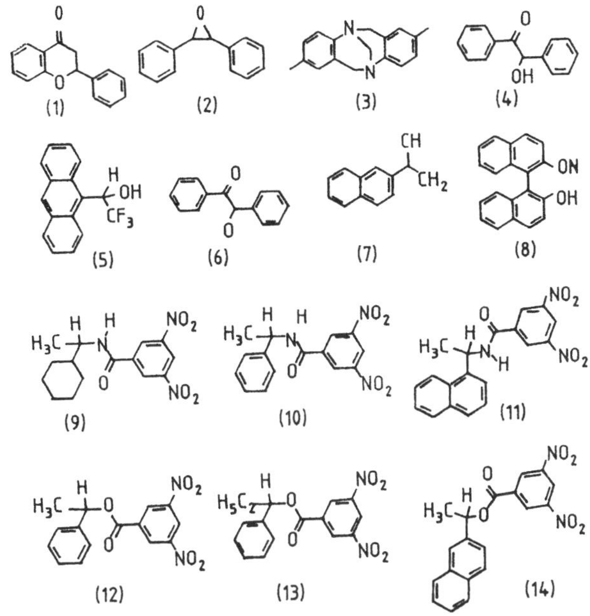

Figure 4.9 *Structure of the racemic solutes*
(Reprinted with permission from ref. 74)

isomers of the industrial intermediates, sodium phenolsulfonates, have been separated on a β-CD-bonded column. It was found that the ionic strength of the mobile phase considerably influenced the efficacy of separation.[79] The chiral selectivity of 3,5-dimethylphenyl carbamoylated α-, β-, and γ-CD derivatives has been compared by using various racemic solutes as model compounds. It was found that the highest selectivity can be achieved under normal-phase conditions.[80] CD-bonded phases have also found application in the analysis of metalloorganic compounds. Thus, β-CD derivatives of cobaltocarboranes of the type 6,6'-μ-R-E(1,7-$C_2B_9H_{10}$)$_2$-2-Co have been efficiently separated on a β-CD column.[81-83]

Cyclodextrins as Mobile Phase Additives

Although chiral stationary phases based on CDs and CD derivatives have been successfully employed for the enantiomeric separation of a wide variety of chiral compounds, each stationary phase is only suitable for the separation of a limited number of chiral compounds. A considerable number of chiral stationary phases are therefore needed for the separation of a range of enantiomeric pairs. The use of CDs as mobile-phase additives can partially or entirely overcome this difficulty.

Separations generally can be carried out on a traditional reversed-phase column by varying the chromatographic conditions, such as the type and concentration of CD, pH, ionic strength and composition of mobile phase *etc.* HPLC methods using CDs as mobile-phase additives are thus versatile and easy to carry out.

Determination of the Strength of Cyclodextrin–Solute Interaction

Reversed-phase HPLC using CDs as mobile-phase additives has been frequently applied for the determination of the thermodynamic parameters of host–guest interactions.

The complex formation constants of imidazole derivatives with β-CD show a high variation indicating that they strongly depend on the chemical structure of the derivatives.[84] As expected they were lower at higher temperatures, suggesting that enthalpy exerts a larger impact on the complex formation than entropy. It was further established that the transfer of solute from the eluent to the stationary phase was enthalpy driven at low concentration of β-CD and entropy driven at higher concentrations.

The thermodynamic parameters of the inclusion complexes of aflatoxins with β-CD have also been determined using a similar RP-HPLC method.[85] Measurements were made on a C_8 column between 30 and 54 °C. Aflatoxins were detected by their fluorescence using isocratic mobile phase of methanol–water (35:65, v/v) containing β-CD in the concentration range of $2.5 \times 10^{-4} - 8 \times 10^{-3}$ M. It was concluded that the interaction between aflatoxins and the stationary phase decreased in the presence of β-CD, probably due to the formation of inclusion complexes.

The effect of the simultaneous application of two CDs in the mobile phase may result in decreased analysis time, and increased enantioselectivity has been observed.[86] Low column temperature also enhanced the efficiency of chiral separation.

The association constants, enthalpies and entropies of complex formation between steroid hormones (estriol, estradiol, ethinyloestradiol and estrone) and β- and γ-CDs have been measured by RP-HPLC using a C_{18} column. The association constant markedly depended on the character of hormone CD pair.[87,88] The quantity and quality of organic modifier in the mobile phase considerably influenced the thermodynamical parameters.[89]

The effect of β-CD on the retention of 17α-estradiol, 17β-estradiol and equilin has been measured at different column temperatures.[90] The results indicated that both the concentration of β-CD and the column temperature influenced the separation. A similar method was employed for the separation of equilin, estrone, 2-, 4- and 16α-hydroxyestrone and hydroxyethyl- and dimethyl-β-CD provided better separation than native β-CD.[91] Fluorescence and UV spectroscopy, RP-HPLC and free solution CE have been applied for the measurement of the formation constant between methyl-β-CD and the oral capillary preservatory drug rutin, 3-[(6-*O*-(6-deoxy-α-L-mannopyranosyl)-β-D-glycopyranosyl)oxy]-2-(3′,4′-dihydroxyphenyl)-5,7-dihydroxy-4*H*-1-benzopyran-4-one. The various methods gave similar but not identical results.[92,93]

Separation of positional and optical isomers of Environmental Pollutants

As the majority of environmental pollutants are volatile or semi-volatile their chiral separation has been generally performed by GC methods and HPLC techniques have found only limited application in this field.

The determination of metabolites of organic solvents in urine samples was performed with RP-PLC using 0, 0.6, 1.0, 1.4 and 1.8% β-CD as mobile-phase additive on a C_{18} column (Figure 4.10).[94] Urine samples from factory workers were injected into the column without any pretreatment. The chromatograms clearly demonstrated that the addition of β-CD not only decreased the retention

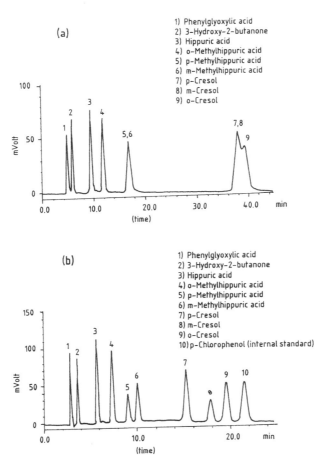

Figure 4.10 *Chromatograms of the nine metabolites (a) in EtOH–H$_2$O–CH$_3$COOH = 20:80:0.3 without β-CD and (b) in the eluent with β-CD and an internal standard (Column: Lichrosorb RP 18 μm, flow rate 0.8 ml min^{-1})* (Reprinted with permission from ref. 94)

time but also enhanced the selectivity of the separation process. These effects may be due to the formation of inclusion complexes.

The complex formation between β-CD and pyrene has been studied in detail by RP-HPLC. The results indicated that the chemical structure of co-modifier exerts a high effect on the stability of the complex.[95] The method has been extended for the study of the complex formation of six polyaromatic hydrocarbons (pyrene, chrysene, fluoranthene, anthracene, benzo[*a*]pyrene, benzo[*ghi*]perylene.[96] The successful separation of fourteen metabolites of benzo[*a*]pyrene using β-[97] and γ-CDs[98] as mobile-phase additives in RP-HPLC has also been reported. Both RP-HPLC and GC have been employed for the study of the complex formation of β-, γ-, and TM-β-CDs with isomeric dimethylnaphthalenes and high differences between the stability of the complexes was observed.[99] Good separation of ochratoxin A and zearalenone in maize has been achieved by using RP-HPLC and β-CD as mobile-phase additive.[100]

Separation of positional and optical isomers of pharmaceuticals

CDs as mobile-phase additives have been frequently used for the analysis of pharmaceuticals, such as steroidal drugs, salicylic acid derivatives, antidepressants, tranquillisers, antitumour agents *etc.* The effect of the addition of β-CD to the mobile phase and that of the column temperature on the separation of steroidal drugs was studied in detail.[101] Retention times of hydrocortisone, testosterone, 17α-methyltestosterone, prednisone, cortisone and 17α-hydroxyprogesterone have been measured on a C_{18} column. The results showed that separation is better at low temperatures and high β-CD concentrations. Optimal conditions were: ACN–water (30:70, v/v) containing 16 mM β-D and 5 °C column temperature.

The impurities of salicylic acid were successfully separated on a phenyl column using β-CD as mobile-phase additive and the results were compared with those obtained on a C_{18} column (Figure 4.11).[102] The chromatograms indicated that the separation was better on the phenyl than on the C_{18} column, and β-CD in the eluent improved separations on both columns. The relationship between peak area and solute concentration was linear between 1 and 100 μg mL^{-1}. Detection limits for salicylic acid, 4-hydroxybenzoic acid, 4-hydroxyisophthalic acid, phenol, gentisic acid, and salicylglycine were 0.07, 0.15, 0.10, 0.12, 0.15 and 0.08 μg mL^{-1}, respectively. The enantiomers of trimipramine (10,11-dihydro-N,N,β-trimethyl-5H-dibenz[*b,f*]azepine-5-propanamine) were extracted with SPE from human serum and were separated with β-CD mobile-phase additive.[103] It was found that 2-hydroxydesmethyl trimipramine and 2-hydroxy trimipramine can be resolved in the eluent ethanol–aqueous 10 mM ammonium acetate buffer pH 4) 5:95 v/v containing 20 mM β-CD. Detector response was linear between 25 and 400 ng mL^{-1} for each enantiomer. The minimum detectable concentration and the limit of quantitation were 10 and 25 ng mL^{-1}, respectively.

The enantiomers of the tranquilliser pentacozine [1,2,3,4,5,6-hexahydro-6,11-dimethyl-3-(3-methyl-2-butenyl)-2,6-methano-3-benzazocine-8-ol] were also

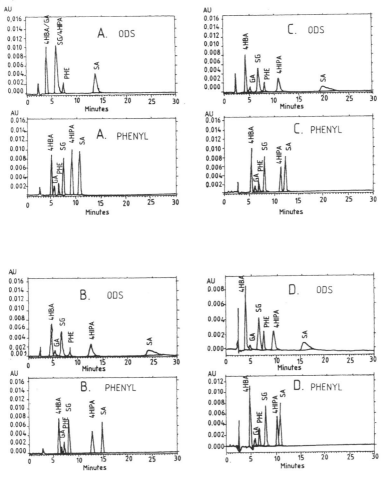

Figure 4.11 *Test mixture eluted with (A) methanol–water–glacial acetic acid, 40:60:1, v/v/v, containing 2.5 g β-CD per litre, pH = 3.0; (B) methanol–water–phosphoric acid, 40:60:1,v/v/v, the pH is adjusted to 2.0 with potassium hydroxide pellets; (C) methanol–water–phosphoric acid, 40:60:1,v/v/v, containing 2.5 g β-CD per litre, the pH is adjusted to 2.0 with potassium hydroxide pellets; (D) methanol–water–phosphoric acid, 40:60:1,v/v/v, containing 5.0 g β-CD per litre, the pH is adjusted to 2.0 with potassium hydroxide pellets. SA = salicylic acid; 4HBA = 4-hydroxybenzoic acid; 4HIPA = 4-hydroxyisophthalic acid; PHE = phenol; GA = gentisic acid; SG = salicylglycine (flow rate 0.8 ml min⁻¹)*
(Reprinted with permission from ref. 102)

separated and quantitatively determined in human serum.[104] Enantiomers were well separated from each other and from the internal standard napthazoline (Figure 4.12). Enantiomers of barbiturates and other bioactive compounds were separated by simultaneously using two different cyclodextrins as mobile phase additivies.[105] Separations were performed on C_{18} columns with water–ethanol mobile phases without CD, with α- or β-CD in the mobile phase, with permethylated CDs in the mobile phase and with native and permethylated CDs in the mobile phase. It was supposed that the good separation capacity of the last system was due to the presence of native CD in the mobile phase and the adsorbed permethylated CD on the surface of the stationary phase. The simultaneous use of two CDs as mobile-phase additives considerably reduced the retention time making possible the decrease of organic modifier in the eluent which results in better chiral discrimination.

The chiral separation capacity of four CDs (β-CD; hydroxypropyl-β-CD, degree of substitution 4.0; methyl-β-CD, degree of substitution 12.7; sulfated-β-CD, degree of substitution 14) was compared using another set of tranquillisers as model compounds.[106] Separations were performed on various RP-HPLC columns coated with C_3, C_6, C_8 and C_{18} alkyl ligands using water–ACN mobile phases. Trebutalin and ibuprofen enantiomers cannot be resolved on the non-porous octadecylsilane column, but terbutalin was separated on hexylsilane column and the type of CD exerted a considerably influences on the efficiency of chiral separation. This result was explained by the supposition that substituents modify the availability of the CD cavity for solutes. It was further established that the enantiomeric resolution of the chromatographic system markedly decreased with increasing concentration of organic modifier in the mobile phase. The isomers of

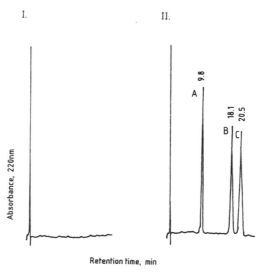

Figure 4.12 *Typical chromatograms of (I) blank serum and (II) serum spiked with 400 ng mL^{-1} of internal standard naphazoline (A) and 75 ng mL^{-1} each of (–)-pentazocine (B) and (+)-pentazocine (C). Column: cyanopropyl with acetonitrile–water ph 5.8*
(Reprinted with permission from ref. 104)

four stereoisomers of the antitumour agent thienyl-5,10-dideazatetrahydrofolate (LY309887) have also been separated by RP-HPLC using CDs as mobile phase additives.[107] The efficacy of nine different HPLC columns, including CD-bonded and protein-bonded chiral stationary phases, C_8, C_{18}, cyano and phenyl phases, was compared. It was established that the best separation was achieved on the phenyl column using β-CD as mobile-phase additive. The results indicated that the concentration of β-CD, pH and ACN equally influences the efficacy of the separation of stereoisomers. The optimisation of the stereoisomer separation of similar antitumour agents (lometrexol sodium and LY231514) was carried out by fractional factorial design using C_{18} and C_1 columns of 25 cm, and β-CD eluent additive.[108]

Terfenadine (a non-sedating H1-receptor antagonist) enantiomers have also been separated by RP-HPLC using β-CD as mobile-phase additives.[109] Resolution of the enantiomers was performed on a C_{18} column with a mobile phase (pH 6.4) consisting of 0.1 M sodium perchlorate buffer–ethanol–diethylamine (70:30: 0.5, v/v) containing 35.24 mM β-CD. The non-linear dependence of $1/k$ on the concentration of β-CD in the mobile phase was explained by the assumption that β-CD forms 2:1 complexes with terfenadine.

Sulfobutyl ether-β-CD (SBE-β-CD) has also been successfully employed for the chiral separation of the calcium channel blocker *rac*-amlodipine (Figure 4.13).[110] It was assumed that the formation of inclusion complexes and ion-pairing interactions between the positively charged enantiomers and the negatively charged SBE-β-CD) accounts for the resolution.

The effect of neutral and charged CD derivatives on the separation of amlopidine enantiomers was compared in HPLC and CE.[111] It was found that

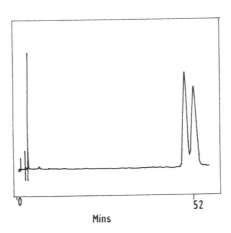

Mins

Figure 4.13 *Optimised separation of amlodipine enantiomers. ACN-potassium dihydrogen phosphate (pH 3.93) containing 2.66 mM SBE-β-CD (26.5:73.5, v/v) at a flow rate of 1.0 mL min^{-1} (detection 238 nm)*
(Reprinted with permission from ref. 110)

neutral CDs was unable to separate the enantiomers of amlopidine, and the separation obtained with CM-β-CD was unsatisfactory in RPLC. In contrast, in CE the enantiomers were successfully separated even with neutral CDs (Figure 4.14). It was established that the S-enantiomer binds more strongly to SBE-β-CD than the R-enantiomer and that the binding occurs in the bulk mobile phase and not on the dynamically coated chiral stationary phase.

CDs have also been employed for the enantiomeric separation of the endogenous neurotoxins salsolinol (1-methyl-6,7-dihydroxy-1,2,3,4-tetrahydroisoquinoline, Sal) and N-methylsalsolinol (NMSal). Human brain and banana samples were extracted with aqueous acid solutions, centrifuged and filtered

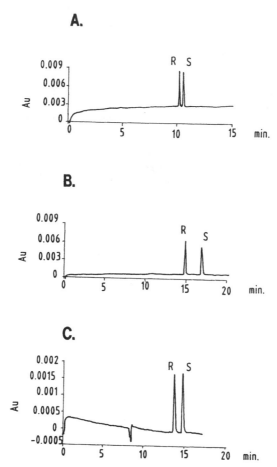

Figure 4.14 *Separation of amlodipine enantiomers in CE using 50 mM NaH$_2$PO$_4$ electrolyte containing (a) 20mM HP-β-CD (pH 3.0) at 20 kV (b) 2.5 mM CM-β-CD (pH 3.0) at 15 kV and (c) 1.0 mM SBE-β-CD (pH 7.0) at 15 kV* (Reprinted with permission from ref. 111)

before HPLC analysis using ion-pair chromatography with β-CD (Figure 4.15).[112] Separation factors depend on the concentration of CD and counter-ion and the differences between the formation constants of the enantiomers.

The inclusion formation constants of the *R*-enantiomers were higher than those of *S*-enantiomers making possible their separation. The separation factor increased non-linearly with increasing concentration of β-CD in the mobile phase. However, higher concentrations (>1 mM) of counter-ion did not influence the enantiomeric separation. The method was validated and employed for the measurement of Sal

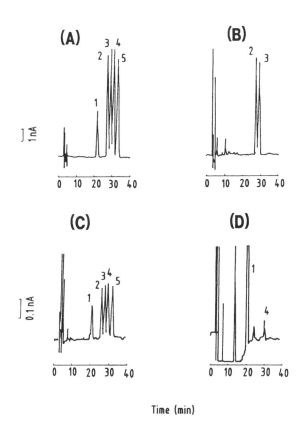

Time (min)

Figure 4.15 *HPLC patterns of samples prepared from dried banana and human brain. (A) Standards, amount per injection for each standard component, 20 pmol; (B) sample prepared from dried banana, (C) standards, amount per injection for each standard component, 0.5 pmol; and (D) sample prepared from human brain grey matter. The mobile phase was 25 mM sodium phosphate buffer pH 3.0, containing 12 mM β-CD, 1 mM sodium heptanesulfonate (SHS), and 0.2% 2-methyl-2-propanol. Peaks: 1, dopamine; 2, (R)-salsolinol; 3, (S)-salsolinol; 4, (R)-N-methylsalsolinol; and 5, (S)-N-methylsalsolinol. (Column: Inertosil ODS-3, flow rate 0.7 ml min⁻¹, eluent; sodium phosphate buffer)* (Reprinted with permission from ref. 112)

and NMSal in human brain grey matter, dried banana and wines.[113] The relationship between the concentration of analytes and the detector response was linear between 0.1 and 40 pm/injection with correlation coefficients 0.96–0.99. Detection limits for dopamine, and R and S-enantiomers of Sal and NMSal were 0.082, 0.047, 0.065, 0.061 and 0.073 pmol/injection, respectively. Recoveries varied between 52.3 and 86.2% and relative standard deviations of intraday and interday reproducibilities were between 7.5 14.1% and 7.0 14.0%, respectively. The method has also been used for the preparation of R-NMSal.[114]

A semi-preparative HPLC technique has been developed and employed for the separation of the enantiomers of the antihistaminic drug brompheniramine using β-CD in the mobile phase as chiral selector. It was stated that the method can also be applied for the semi-preparative isolation of other enantiomer pairs.[115] Carboxymethyl-β-CD derivatives with various degrees of substitution have been used as mobile-phase additives for the chiral separation of some basic β-CD drugs (oxprenolol, ephedrine). The results indicated that the degree of substitution exerts a considerable influence on the efficacy of chiral resolution.[116]

Separation of positional and optical isomers of Miscellaneous Organic Compounds

The effect of α-, β- and γ-CDs on the enantiomeric separation of alanine-β-naphthylamide (Ala-β-NA) and 1-(1-naphthyl)ethylamine (1-NEA) was studied in detail.[117] Separations were performed on a CROWNPAK CR(+) column (C_{18} support dynamically coated with a hydrophobic chiral crown ether) at 40 °C. The effect of pH and ionic strength of the mobile phase on the efficacy of separation was determined. The results indicated that α- and γ-CDs do not exert a marked influence either on the separation factor or on the retention factors, which decreased with increasing concentrations of β-CD in the mobile phase without deterioration of resolution. This result was explained by the formation of inclusion complexes between the solutes and β-CD. Retention factors were lower in the presence of ions, K^+ being the most effective. The effect of pH on the enantiomeric separation was of secondary importance. It was stated that the method is suitable for the detection of 0.05% of the D-form in L-Ala-β-NA: therefore it can be used as a quality control method.

Starch branching enzyme I (SBE-I) was also isolated from potato by γ-CD affinity chromatography.[118] A typical chromatogram is shown in Figure 4.16. The method was proposed for the isolation of other soluble starch branching enzymes with different substrate specificities.

The performance of HPLC methods using CDs was compared with that of TLC. The comparison of various methods promotes the better understanding of the theoretical background of separation and facilitates the design of optimal separation process. The inclusion complexes of phenolic and cinnamic acids with β-CD were studied by the simultaneous application of TLC and HPLC.[119] Silanised silica and C_2 plates were employed for TLC using ACN–aqueous buffer mixtures as the mobile phase. HPLC separation was carried out on a C_{18} column

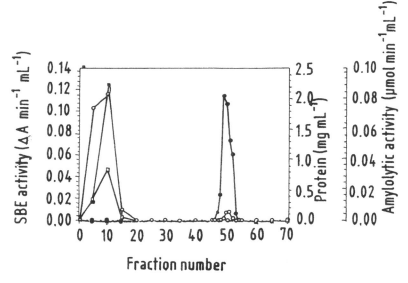

Figure 4.16 *γ-Cyclodextrin affinity chromatography of the potato starch branching enzyme. Protein was eluted with 5 mM γ-cyclodextrin, beginning at fraction forty. Protein, ○; starch branching enzyme ●; α-amylase, ■; β-amylase, □ (Column: γ-cyclodextrin with aqueous eluent)*
(Reprinted with permission from ref. 118)

using a similar eluent system. The separation factors clearly show that the separation of *cis/trans*-cinnamic acids was improved with increasing concentration of β-CD in the mobile phase (Table 4.2). A linear relationship ($r = 0.951$) was found between the R_M values determined in RP-TLC and the retention factors (log k). by RP-TLC. Thus RP-TLC could be used as a pilot method for HP.

D-Alanine and D-glutamic acid have been measured in biological samples by coupled-column chromatography using β-CD as chiral selector.[120] Enantiomers of amino acids have been separated after derivatisation on a C_{18} column using β-CD and heptakis-(2,6-di-O-methyl)-β-CD. It was found that the derivatised CD shows a higher enantioselectivity.[121] The enantioselectivities of β-CD, carboxymethyl-β-CD and a cationic β-CD have been compared using phenylhydantoin and methylhydantoin amino acids. The advantageous characteristics of the cationic derivative was established.[122] The successful separation of sulfates of glucuronides of provitamin D, vitamin D, and 25-hydroxyvitamin D_3 has also been reported. The method employed RP-HPLC and heptakis-(2,6-di-O-methyl)-β-CD as a mobile-phase additive.[123] Similar techniques have been used for the separation of vitamin D_2 D_5, and provitamin D_2 D_5.[124]

Table 4.2 *HPLC selectivity factors (α) for cis-trans cinnamic acids with and without β-CD in the mobile phase (30% methanol in 0.04 M phosphate buffer solution (pH 3.0))*

Compound	Without β-CD	0.002 M β-CD	0.005 M β-CD	0.007 M β-CD	0.01 M β-CD
Cinnamic acid	–	1.08	1.06	1.10	1.36
o-Coumaric acid	1.10	1.30	1.35	1.50	1.50
m-Coumaric acid	1.15	1.36	1.30	1.59	1.92
p-Coumaric acid	1.07	1.44	1.25	1.34	1.33
Caffeic acid	1.06	1.50	1.67	2.50	3.00
Ferulic acid	1.12	1.33	1.43	1.55	2.14
Isoferulic acid	1.07	1.26	1.31	1.21	1.25

(Reprinted with permission from ref. 119)

References

1. T. Momose, M. Mure, T. Iida, J. Goto and T. Nambara, *J. Chromatogr. A,* 1998, **811**, 171–180.
2. I. R. Politzer, K. T. Crago, T. Hollin and M. Young, *J. Chromatogr. Sci.,* 1995, **33**, 316–320.
3. P. K. Zarzycki, M. Wierzbowska, J. Nowakowska, A. Chmielewska and H. Lamparczyk, *J. Chromatogr. A,* 1999, **839**, 149–156.
4. T. Cserháti, E. Forgács and J. Holló, *J. Pharm. Biomed. Anal.,* 1995, **13**, 533–541.
5. T. Cserháti, *Anal. Biochem.,* 1995, **225**, 328–332.
6. T. Cserháti, *Int. J. Pharm.,* 1995, **124**, 205–211.
7. Y. Darwish and T. Cserháti, *Toxicol. Environ. Chem.,* 1996, **54**, 139–147.
8. T. Cserháti, E. Forgács and J. Szejtli, *Int. J. Pharm.,* 1996, **141**, 1–7.
9. E. Forgács and K. Demnerová, *Biomed. Chromatogr.,* 1996, **10**, 92–94.
10. T. Cserháti and E. Forgács, *Anal. Biochem.,* 1998, **246**, 205–210.
11. T. Cserháti and E. Forgács, *Chemom. Intell. Lab. Syst.,* 1998, **40**, 93–100.
12. T. Cserháti and E. Forgács, *Eur. J. Pharm. Biopharm.,* 1998, **46**, 153–159.
13. T. Cserháti and E. Forgács, *J. Pharm. Biomed. Anal.,* 1998, **18**, 179–185.
14. T. Cserháti and E. Forgács, *Carbohydr. Polym.,* 1999, **38**, 171–177.
15. E. Forgács and T. Cserháti, *J. Chromatogr. A,* 1999, **845**, 447–453.
16. T. Cserháti, G. Csiktusnádi Kiss and J. Augustin, *J. Incl. Phenom. Macrocycl. Chem.,* 1999, **33**, 123–133.
17. T. Cserháti, E. Forgács and Gy. Sági, *J. Liq. Chromatogr. Relat. Technol.,* 1999, **22**, 125–135.
18. L. Lepri, *J. Planar Chromatogr. Mod. TLC,* 1997, **10**, 320–331.
19. V. Lambroussi, S. Piperaki and A. Tsantili-Kakoulidou, *J. Planar Chromatogr. Mod. TLC,* 1999, **12**, 124–128.

20. M.-B. Huang, H.-K. Li, G.-L. Li, C.-T. Yan and L.-P. Wang, *J. Chromatogr. A,* 1996, **742**, 289–294.
21. S. M. Han and D. W. Armstrong, in *Chiral Separations by HPLC Applications to Pharmaceutical Compounds*, ed. A. M. Krstulovic, Ellis Harwood 1989, Chichester, pp. 208–286.
22. F. Bressolle, M. Audran, T.-N. Pham and J.-J. Vallon, *J. Chromatogr. B,* 1996, **687**, 303–336.
23. N. Morin, Y. C. Guiilaume, E. Peyrin and J.-C. Rouland, *Anal. Chem.,* 1998, **70**, 2819–2826.
24. P. D. Ferguson, D. M. Goodall and J. S. Loran, *J. Chromatogr. A*, 1997, **768**, 29–38.
25. K. Weber, R. Kreuzig and M. Bahadir, *Chemosphere,* 1997, **35**, 13–20.
26. P. Haglund, *J. Chromatogr. A,* 1996, **724**, 219–228.
27. P. Haglund, *Chemosphere*, 1996, **32**, 2133–2140.
28. M. C. Ringo and C. E. Evans, *Anal. Chem.,* 1997, **69**, 643–649.
29. D. W. Armstrong, X. Wang, L. W. Chang, H. Ibrahim, G. R. Reid III and T. E. Beesly, *J. Liq. Chromatogr. Relat. Technol.,* 1997, **20**, 3297–3308.
30. G. Crini, M. Morcellet and G. Torri, *J. Chromatogr. Sci.,* 1996, **34**, 477–484.
31. G. Crini and M. Morcellet, *J. Chromatogr. Sci.,* 1996, **34**, 485–494.
32. G. Crini, Y. Lekchiri and M. Morcellet, *Chromatographia*, 1993, **40**, 296–302.
33. M. D. Müller and H.-R. Buser, *Anal. Chem.*, 1994, **66**, 2155–2162.
34. M. D. Müller and H.-R. Buser, *Anal. Chem.*, 1995, **67**, 2691–2698.
35. R. Furuta and H. Nakazawa, *J. Chromatogr.*, 1992, **625**, 231–235.
36. S. Piperaki, A. Perakis and M. Parissi-Poulou, *J. Chromatogr. A,* 1994, **660**, 339–350.
37. G. W. Ponder, S. L. Butram, A. G. Adams, C. S. Ramanathan and J. T. Stewart, *J. Chromatogr. A*, 1995, **692**, 173–182.
38. C. Pham-Huy, N. Chikhi-Chorfi, H. Galons, N. Sadeg, X. Laqueille, N. Aymard, F. Massicot and J.-M. Warnet, *J. Chromatogr. B,* 1997, **700**, 155–163.
39. R. L. G. Norris, P. J. Ravenscroft and S. M. Pond, *J. Chromatogr. B,* 1994, **661**, 346–350.
40. G. Ecker, E. Mohr, R. Geyer and W. Fleischhacker, *Sci. Pharm.,* 1996, **64**, 1–11.
41. F. Sadeghipour and J.-L. Veuthey, *Chromatographia,* 1998, **47**, 285–290.
42. A. M. Rizzi, R. Hirz, S. Cladrowa-Runge and H. Jonsson, *Chromatographia*, 1994, **39**, 131–137.
43. S. Piperaki and M. Parissi-Poulou, *Chirality,* 1993, **5**, 258–266.
44. S. Piperaki, S. G. Penn and D. M. Goodall, *J. Chromatogr. A,* 1995, **700**, 59–67.
45. M. C. Ringo and C. E. Evans, *Anal. Chem.,* 1997, **69**, 4964–4971.
46. C. Pham-Huy, B. Radenen, A. Sahui-Gnassi and J.-R. Claude, *J. Chromatogr. B*, 1995, **665**, 125–132.

47. S. C. Chang, G. L. Reid III, S. Chen, C. D. Chang and D. W. Armstrong, *TRAC*, 1993, **12**, 144–153.

48. A. Burmester and B. Jastorff, *J. Chromatogr. A*, 1996, **749**, 25–32.

49. J. Cizmarik, J. Lehotay, K. Hromuláková, M. Pokorná and M. Lacuska, *Pharmazie*, 1997, **52**, 402.

50. N. Thuaud and B. Sebille, *J. Chromatogr. A*, 1994, **685**, 15–20.

51. T. A. Walker, *J. Chromatogr.*, 1993, **633**, 97–103.

52. E. Tesarová, M. Gilar, P. Hobza, M. Kabelác, Z. Deyl and E. Smolková-Keulemansová, *J. High Resolut. Chromatogr.*, 1995, **18**, 597–601.

53. M. D. Beeson and G. Vigh, *J. Chromatogr.*, 1993, **634**, 197–204.

54. A. M. Stalcup and K. H. Gahm, *Anal. Chem.*, 1996, **68**, 1369–1374.

55. G. Félix, C. Cachau, A. Thienpont and M.-H. Soulard, *Chromatographia*, 1996, **42**, 583–590.

56. H. Nishi, K. Nakamura, H. Nakai, T. Sato and S. Terabe, *Chromatographia*, 1995, **40**, 638–644.

57. M. Schulte, R. Ditz, R. M. Devant, J. N. Kinkel and F. Charton, *J. Chromatogr. A*, 1997, **769**, 93–100.

58. R. H. Pullen, J. J. Brennan and G. Patonay, *J. Chromatogr. A*, 1995, **691**, 187–193.

59. J. Breinholt, S. V. Lehmann and A. R. Varming, *Chirality*, 1999, **11**, 768–771.

60. Y. Tang, J. Zukowski and D. W. Armstrong, *J. Chromatogr. A*, 1996, **743**, 261–271.

61. T. Iida, H. Matsunaga, T. Fukushima, T. Santa, H. Homma and K. Imai, *Anal. Chem.*, 1997, **69**, 4463–4468.

62. J. W. Ryu, D. W. Kim, K.-P. Lee, D. Pyo and J. H. Park, *J. Chromatogr. A*, 1998, **1998**, 247–252.

63. T. Araki, Y. Kashiwamoto, S. Tsunoi and M. Tanaka, *J. Chromatogr. A*, 1999, **845**, 455–462.

64. B. Sébille, M. Guillaume, C. Vidal-Madjar and N. Thuaud, *Chromatographia*, 1997, **45**, 383–389.

65. D. W. Armstrong, L. W. Chang, S. C. Chang, X. Wang, H. Ibrahim, G. R. Reid III and T. E. Beesley, *J. Liq. Chromatogr. Relat. Technol.*, 1997, **20**, 3279–3295.

66. W. L. Hinze, T. E. Riehl, D. W. Armstrong, W. DeMond and T. Ward, *Anal. Chem.*, 1985, **57**, 237–242.

67. M. Tanaka, M. Yoshinaga, M. Ito and H. Ueda, *Anal. Sci.*, 1995, **11**, 227–231.

68. A. M. Rizzi, S. Cladrowa-Runge, H. Jonsson and S. Osla, *J. Chromatogr. A*, 1995, **710**, 287–295.

69. M. Pawlowska, S. Chen and D. W. Armstrong, *J. Chromatogr.*, 1993, **641**, 257–265.

70. S. Chen, M. Pawlowska and D. W. Armstrong, *J. Liq. Chromatogr.*, 1994, **17**, 483–497.

71. J. Zukowski, M. Pawlowska, M. Nagatkina and D. W. Armstrong, *J. Chromatogr.*, 1993, **629**, 169–179.

72. B. Grüner, J. Holub, J. Plesek, T. Vanek and H. Votarová, *J. Chromatogr. A,* 1998, **793**, 249–256.

73. J. Feurle, H. Jomaa, M. Wilhelm, B. Gutsche and M. Herderich, *J. Chromatogr. A,* 1998, **803**, 111–119.

74. C. Cachau, A. Thienpont, M.-H. Soulard and G. Félix, *Chromatographia,* 1997, **44**, 411–416.

75. S. L. Abidi and T. L. Mounts, *J. Chromatogr. A,* 1994, **670**, 67–75.

76. P. J. Simms, A. T. Hotchkiss, Jr., P. L. Irwin and K. B. Hicks, *Carbohydr. Res.,* 1995, **278**, 19.

77. D. Schumacher and L. W. Kroh, *Food Chem.,* 1995, **54**, 353–356.

78. Y. Deng, W. Maruyama, P. Dostert, T. Takahashi, M. Kawai and M. Naoi, *J. Chromatogr. B,* 1995, **670**, 47–54.

79. D. R. Wilder, G. W. Tindall, L. J. Cunningham and J. L. Little, *J. Chromatogr.,* 1993, **635**, 221–226.

80. T. Hargitai and Y. Okamoto, *J. Liq. Chromatogr.,* 1993, **16**, 843–858.

81. J. Plesek, B. Grüner, T. Vanek and H. Votavová, *J. Chromatogr.,* 1993, **633**, 73–80.

82. J. Plesek, B. Grüner, J. Fusek and H. Votavová, *Collect. Czech. Chem. Commun.,* 1993, **58**, 2936–2943.

83. J. Plesek, B. Grüner, S. Hermánek, J. Fusek and H. Votavová, *Collect. Czech. Chem. Commun.,* 1994, **59**, 374–380.

84. N. Morin, Y. C. Guillaume, E. Peyrin and J.-C. Rouland, *J. Chromatogr. A,* 1998, **808**, 51–60.

85. C. M. Franco, C. A. Fente, B. I. Vázquez, A. Cepeda, G. Mahuzier and P. Prognon, *J. Chromatogr. A,* 1998, **815**, 21–29.

86. A. Bieljewska, R. Nowakowski, K. Duszczyk and D. Sybilska, *J. Chromatogr. A,* 1999, **840**, 159–170.

87. N. Sadlej-Sosnowska, *J. Pharm. Biomed. Anal.,* 1995, **13**, 701–704.

88. N. Sadlej-Sosnowska, *Eur. J. Pharm. Sci.,* 1995, **3**, 1–5.

89. N. Sadlej-Sosnowska, *J. Chromatogr. A,* 1996, **728**, 89–95.

90. H. Lamparczyk and P. K. Zarzycki, *J. Pharm. Biomed. Anal.,* 1995, **13**, 543–549.

91. B. J. Spencer and W. C. Purdy, *J. Liq. Chromatogr.,* 1995, **18**, 4063–4080.

92. S. Letellier, B. Maupas, J. P. Gramond, F. Guyon and P. Gareil, *Anal. Chim. Acta,* 1995, **315**, 357–363.

93. Y. L. Loukas, E. Antoniadou-Vyza, A. Papadaki-Valiraki and K. G. Machera, *J. Agric. Food Chem.,* 1994, **42**, 944–948.

94. D. H. Moon, N. W. Paik and Y.-B. Shim, *J. Chromatogr. B,* 1997, **694**, 367–374.

95. N. Husain, V. C. Anigbogu, M. R. Cohen and I. M. Wartner, *J. Chromatogr.,* 1993, **635**, 211–219.

96. N. Husain, A. Y. Cristian and I. M. Warner, *J. Chromatogr. A,* 1995, **699**, 73–83.

97. M. Rozbeh and R. J. Hurtubise, *J. Liq. Chromatogr.,* 1995, **18**, 17–37.

98. M. Rozbeh and R. J. Hurtubise, *J. Liq. Chromatogr.,* 1995, **18**, 1909–1931.

99. D. Sybilska, M. Asztemborska, A. Bielejewska, J. Kowalczyk, H. Dodziuk,

K. Duszczyk, H. Lamparczyk and P. Zarzycki, *Chromatographia*, 1993, **35**, 637–642.

100. V. Seidel, E. Poglits, K. Schiller and W. Lindner, *J. Chromatogr.*, 1993, **635**, 227–235.

101. P. K. Zarzycki, M. Wierzbowska and H. Lamparczyk, *J. Pharm. Biomed. Anal.*, 1996, **14**, 1305–1311.

102. J. D. Goss, *J. Chromatogr. A*, 1998, **828**, 267–271.

103. E. Ameyibor and J. T. Stewart, *J. Liq. Chromatogr. Relat. Technol.*, 1997, **20**, 3107–3119.

104. E. Ameyibor and J. T. Stewart, *J. Chromatogr. B*, 1997, **703**, 273–278.

105. R. Nowakowski, A. Bielejewska, K. Duszczyk and D. Sybilska, *J. Chromatogr. A*, 1997, **782**, 1–11.

106. E. Ameyibor and J. T. Stewart, *J. Liq. Chromatogr. Relat. Technol.*, 1997, **20**, 855–869.

107. D. W. Marks, *J. Chromatogr. Sci.*, 1997, **35**, 201–205.

108. L. M. Osborne and T. W. Miyakawa, *J. Liq. Chromatogr. Relat. Technol.*, 1997, **20**, 501–509.

109. S. Surapaneni and S. K. W. Khalil, *J. Pharm. Biomed. Anal.*, 1996, **14**, 1631–1634.

110. P. K. Ovens, A. F. Fell, M. W. Coleman and J. C. Berridge, *Chirality*, 1996, **8**, 466–476.

111. P. K. Owens, A. F. Fell, M. W. Coleman and J. C. Berridge, *J. Chromatogr. A*, 1998, **797**, 187–195.

112. Y. Deng, W. Maruyama, H. Yamamura, M. Kawai, P. Dostert and M. Naoi, *Anal. Chem.*, 1996, **68**, 2826–2931.

113. Y. Deng, W. Maruyama, M. Kawai, P. Dostert, H. Yamamura, T. Takahashi and M. Naoi, *J. Chromatogr. B*, 1997, **689**, 313–320.

114. C. Minami, Y. Deng, F. Urano, T. Kaiya, K. Kohda, M. Kawai and M. Naoi, *J. Chromatogr. B*, 1997, **702**, 245–248.

115. A. D. Cooper and T. M. Jefferies, *J. Chromatogr.*, 1993, **637**, 137–143.

116. J. Szemán, K. Ganzler, A. Salgó and J. Szejtli, *J. Chromatogr. A*, 1996, **728**, 423–431.

117. Y. Mashida, H. Nishi and K. Nakamura, *J. Chromatogr. A*, 1999, **830**, 311–320.

118. A. Vikso-Nielsen and A. Blennow, *J. Chromatogr. A*, 1998, **800**, 382–385.

119. M. L. Bieganowska, A. Petrucznik and K. Glowniak, *J. Planar Chromatogr. Mod. TLC*, 1995, **8**, 63–68.

120. A. M. Rizzi, P. Briza and M. Breitenbach, *J. Chromatogr.*, 1992, **582**, 35–40.

121. B. J. Spencer and W. C. Purdy, *Anal. Lett.*, 1995, **28**, 1865–1881.

122. C. Roussel and A. Favrou, *J. Chromatogr. A*, 1995, **704**, 67–74.

123. K. Shimada, K. Mitamura, H. Kaji and M. Morita, *J. Chromatogr. Sci.*, 1994, **32**, 107–111.

124. K. Shimada, K. Mitamura, M. Miura and A. Miyamoto, *J. Liq. Chromatogr.*, 1995, **18**, 2885–2893.

Use of Cyclodextrins in Electrophoretic Techniques

Because of the high separation efficiency and very small injection volume, modern capillary electrophoretic techniques, such as capillaryelectrophoresis (CE), isotachophoresis, capillary gel electrophoresis (CGE), micellar electrokinetic chromatography (MEKC) *etc.*, have become well established analytical methods not only for the separation of large biomolecules but also for the analysis of small solutes. The theory and practice of enantiomeric separation by electrophoretic techniques has been reviewed.[1-4] Separate reviews have been devoted to the application of capillary electrophoretic methods for the chiral separation of pharmaceuticals.[5-7]

1 Capillary Zone Electrophoresis

The high theoretical plate numbers, ruggedness, relatively short separation time and versatility make CE methods valuable alternatives to other chromatographic techniques (HPLC, GC and SFC) for the separation of enantiomers. The use of CDs and CD derivatives in CE has been frequently reviewed. Thus, the employment of charged CDs as chiral selectors,[8,9] the influence of the type and concentration of the chiral selector, ionic strength, pH, composition of background electrolyte and organic modifier on the efficiency of enantiomeric separation,[10] the use of mixed CDs,[11] the application of CDs for the enhancement of efficiency of achiral separations[12] and the characterisation of inclusion complexes[13] have been discussed in detail.

Separation of Positional and Optical Isomers of Pesticides

The widespread application of pesticides may result in increased pollution of surface and ground waters and soil. The importance of their separation and quantitative determination in complicated organic and inorganic matrices at very low concentrations promoted the study of the employment of CE methods for their analysis. Thus, the separation of chlorophenoxy acid herbicides by

CD-modified capillary CE has been reported.[14] Electrophoresis buffers were aqueous solutions of 0.1 M disodium hydrogen phosphate and sodium dihydrogen phosphate and the herbicides were detected at 200 nm. The herbicides were not separated efficiently in CD-free buffers but their separation was achieved in buffers containing α or β-CD (Figure 5.1). It was further established that the simultaneous use of 4 mM of α- and 1 mM of β-CD is suitable for the separation of each herbicide and the enantiomers of 2,4-DP and 2,4,5-TP. The formation of inclusion complexes depended considerably on the steric correspondence between herbicides and CDs. Recoveries of chlorophenoxy acid herbicides from lake water varied between 73% and 90% and were unaffected by the accompanying matrix.

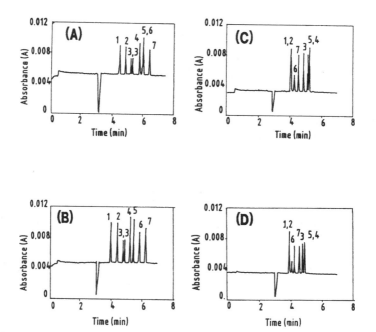

Figure 5.1 *Electropherogram of the chlorophenoxy acid herbicides in buffer with different cyclodextrin concentrations. (A) 1 mM α-CD; (B) 2 mM α-CD. Voltage, 25 kV; separation column, 47 cm long (40 cm to detector) × 50 μm ID; buffer, phosphate buffer pH 5.6; conductivity, 2.05 mS cm⁻¹; pressure injection, 4 s; detector wavelength 200 nm. (C) 2 mM β-CD; (D) 7 mM β-CD. Conductivity, 1.74 mS cm⁻¹; others are the same as for α-CD. Peak identification; 1 = 4-(2,4-dichlorophenoxy)butanoic acid (2,4-DB); 2 = 4-(4-chloro-2-methylphenoxy)butanoic acid (MCPB); 3 = 2-(2,4-dichlorophen-oxy)propionic acid (2,4-DP); 4 = 2,4-dichlorophenoxyacetic acid (2,4-D); 5 = 4-chloro-2-methyl-phenoxyacetic acid (MCPA); 6 = 2-(2,4,5-trichlorophenoxy)propionic acid (2,4,5-TP), and 7 = 2,4,5-trichlorophenoxyacetic acid (2,4,5-T)*
(Reprinted with permission from ref. 14)

A slightly different CE method was developed and applied for the separation of another set of phenoxy acid herbicides.[15] Separations were carried out in an uncoated fused-silica capillary. Separation buffers consisted of 0.1 M of sodium borate 0.05 M sodium phosphate buffers (pH 9.0) containing CDs (Figure 5.2).

Another study was dedicated to the enantiomeric separation of 2,4-DP isomers using new CD derivatives.[16] Separations were performed in a polyacrylamide-coated capillary with a running electrolyte consisting of 100 mM acetic acid sodium acetate buffer (pH 5.0) with a trace of potassium bromide. The inclusion constants of 2,4-DP with CDs (K_{A-CD}) were calculated from the dependence of the corrected electrophoretic mobility of the enantiomer (μ_A) by

$$\mu_A = (\mu_f + \mu_{compl} K_{A-CD}[CD])/(1 + K_{A-CD}[CD]) \tag{5.1}$$

where μ_f is the electrophoretic mobility of the free enantiomer, μ_{compl} is the electrophoretic mobility of the complexed enantiomer and [CD] is the concentration of CD. It was found that the concentration and type of CD equally influence the efficiency of chiral separation. Molecular modelling suggested that the guest molecule fits tightly into the cavity of β-CD resulting in high energy of

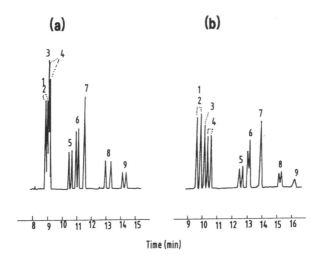

(a) **(b)**

Time (min)

Figure 5.2 *Electropherograms of phenoxy acid herbicides with (a) 5 mM β-CD and 10 mM 2,3-dimethyl-α-CD and (b) 2.5 mM of each α-CD and 2,3-dimethyl-α-CD. Peaks; 1 = 2,4-DP; 2 = 2-(4-chloro-2-methylphenoxy)propionic acid (Meco-prop); 3 = 2,4-D; 4 = 2-(4-chlorophenoxy)propionic acid (2,4-CPPA); 5 = 2-(3-chlorophenoxy)propionic acid (2,3-CPPA); 6 = 2-(2,4,5-trichlorophenoxy)-propionic acid (Silvex); 7 = 2,4,5-T; 8 = 2-(2-chlorophenoxy)propionic acid (2,2-CPPA), and 9 = 2-phenoxypropionic acid (2-PPA) with uncoated fused-silica capillary, constant voltage 20 kV (Reprinted with permission from ref. 15)*

interaction, while larger CDs allow the guest molecule to find a position with minimal energy. Substituents on the CDs hinder the entrance of the guest molecule into the cavity but reduce its degree of freedom. It was emphasised that chiral separation depended on the difference between the strength of inclusion complexes of enantiomers and not on the absolute value of the inclusion constant. Another study established that the enantiomeric separation of phenoxy acid herbicides depended on the type of CD used as chiral selector.[17] The chiral separation capacities of native α-, β-, γ-, DM-β-and TM-β-CD have been compared in the CE analysis of dichlorprop, mecoprop and fenoprop. The data indicated that the best separation can be obtained with 25 mM TM-β-CD in 50 mM acetate buffer.[18] The derivatisation of phenoxy acid herbicides with 7-aminonaphthalene-1,3-disulfonic acid and fluorescence detection resulted in a detection limit of 0.2 ppb. Moreover, the chiral separation of derivatives was better than that of underivatised herbicides in the presence of TM-β-CD.[19] The effect of native CDs on the separation of nine plant growth regulators by CE has also been reported.[20]

Separation of Positional and Optical Isomers of Other Environmental Pollutants

Nitrophenols are by-products of many industrial processes, such as dye, pesticide and explosive production, and cause serious environmental pollution. A CE method using polyvinylpyrrolidone (PVP) and β-CD as mobile-phase additives has been developed for their separation (Figure 5.3) and quantitative determination and the results were compared with those obtained by isotachophoresis (ITP).[21] The electropherograms show that the best separation can be achieved by using PVP as a buffer additive. The day-to-day reproducibility of the migration times was lower than 1%, and the detection limits were between 20 and 80 ppb. Reproducibility of the determination of nitrophenols varied from 1 to 5% in the 1–6 ppm concentration range.

Nitrophenols and phenylbutyrates have also been separated by CE using α-CD as buffer additive.[22] Mesityl oxide was used as electroosmotic flow marker. Polycyclic aromatic hydrocarbons have also been separated by CD-modified CE.[23] Priority PAHs were adsorbed by a glass fibre from diluted aqueous samples. After equilibrium the glass fibre was connected to the separation capillary *via* a special adapter. Separations were carried out at 30 kV using 50 mM borate buffer (pH 9.2) containing CDs in different concentrations (Figure 5.4). It was established that the electropherograms did not change after 2 h equilibration time. The sensitivity of the methods varied between 8 ppb (pyrene) and 75 ppb (acenaphthene). It was stated that the SPME method is suitable for the preconcentration and cleaning of samples. Because of the good separation characteristics and high sensitivity the method was proposed for the analysis of PAHs in soils and water. Native β-CD has also been used for the CE separation of PAHs.[24] The separation of five selected PAHs (naphthalene, anthracene, phenanthrene, fluoranthene, pyrene) was carried out in a fused silica capillary at 20 kV applied

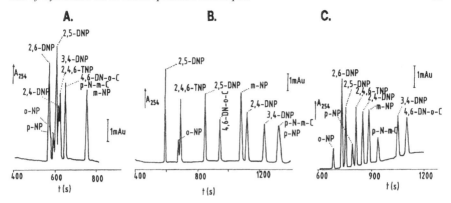

Figure 5.3 *CZE separation of nitrophenols without additives (A), with β-CD (B) and with PVP (C). Driving current, 100 μA, the concentrations of the analytes in a model mixture were 5–10 ppm. Peak identification; o-NP = o-nitrophenol; p-NP = p-nitrophenol; 2,4-DNP = 2,4-dinitrophenol; 2,5-DNP = 2,5-dinitrophenol; 2,6-DNP = 2,6-dinitrophenol; 3,4-DNP = 3,4-dinitrophenol; 2,4,6-TNP = picric acid; p-M-m-C = p-nitro-m-cresol, and 4,6-DN-o-C = 4,6-dinitro-o-cresol with 300 μm ID (650 μm OD) capillary tube made of fluorinated ethylene-propylene copolymer, detection 254 nm*
(Reprinted with permission from ref. 21)

voltage. Samples were injected hydrostatically and the PAHs were detected at 254 nm. The running buffer consisted of water–methanol 90:10 (v/v), pH 12.9 containing 7 M urea and 10 mM β-CD. It was established that the pH exerted a decisive role. The dependence of the resolution on the concentration of β-CD in the running buffer showed marked optimum.

It was further established that the application of mixtures of neutral and anionic β-CDs considerably enhanced the selectivity of CE separation of PAHs.[25] The method has been employed for the determination of PAHs in contaminated soils. Another study used the SFE extraction of PAHs from soils and the higher separation capacity of CE was compared with that of HPLC.[26]

The enantiomers of 3,4-dihydro-2*H*-1-benzopyran derivatives have been separated by CE using β-CD as chiral selector. It was found that enantioselectivity depended non-linearly on the concentration of β-CD in the buffer.[27]

The enantiomers of warfarin have been separated and quantitatively determined in human plasma by CE using methylated β-CD as buffer additive. The detection limit of the method was 0.2 mg mL^{-1}.[28] Warfarin and other acidic and basic enantiomers have been separated in CE by 6^A-methylamino-β-CD and heptamethylamino-β-CD. It was established again that the chiral separation always depended on the character of CD analyte pair.[29]

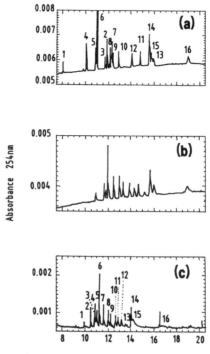

Figure 5.4 *Electropherograms of PAHs obtained with 50 mM borate buffer pH 9.2. (a) 35 mM sulfobutyloxy-β-CD (SBβCD), 15 mM methyl-β-CD (MβCD), sample 20x. (b) 70 mM SBβCD, sample 40x. (c) 35 mM SBβCD, 15 mM MβCD, 4 mM β-CD, sample 40x. Peak identification; 1 = di-benz[a,h]anthracene; 2 = acenaphthylene; 3 = acenaphthene; 4 = naphthalene; 5 = fluorene; 6 = anthracene; 7 = phenanthrene; 8 = chrysene; 9 = benz[a]anthracene; 10 = benzo[k] fluoranthene; 11 = fluoranthene; 12 = benz[a]pyrene; 13 = pyrene; 14 = benzo [b]-fluoranthene; 15 = indeno[1,2,3-cd]pyrene, and 16 = benzo[g,h,i]perylene with P/ACE 5000 Beckman Fullerton inst.*
(Reprinted with permission from ref. 23)

Separation of Positional and Optical Isomers of Amino Acids and Related Compounds

Because of their outstanding biological importance the enantiomers of amino acids and peptides have been separated by various CE techniques using CD additives. In order to increase the separation efficiency and sensitivity, amino acids have been generally derivatised. Thus, *N-tert*-butoxycarbonyl (Boc) amino acids were successfully separated by CE using hydroxypropyl-β-CD (HP-β-CD) in the running buffer.[30] The *N-t*-Boc amino acids were detected at 200 nm. The running buffers consisted of 50 mM phosphate (pH 7.0) containing of methanol and HP-β-CD. It was established that the efficiency of enantiomeric separation increased with increasing concentration of methanol in the running buffer. However, higher concentrations (25 or 40 vol%) had no additional beneficial effect on the separation. Interestingly, β-CD was not able to separate the enantiomers the optimal HP-β-CD concentration was 5–15 mM. It was assumed that the hydrophobic *N-tert*-butoxycarbonyl substructure is included in the cavity of HP-β-CD, and the chiral centre is bound to the outer surface of the CD molecule. These results suggest the involvement of hydrophilic forces in the interaction.

Racemic amino acids derivatised with 9-fluoromethyl chloroformate (Fmoc) have also been separated by CE using β- and γ-CDs as chiral selectors (Figure 5.5).[31] The background electrolyte in each instance was 50 mM phosphate buffer. It was found that the concentrations of both sodium dodecylsulfate (SDS) and CD exerted a marked effect on the efficiency of the enantiomeric separation, β-CD

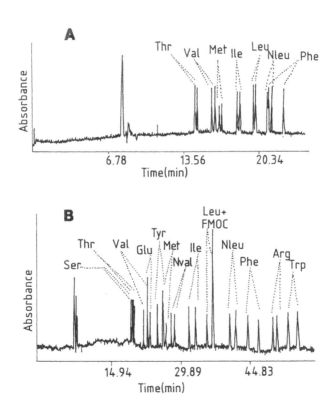

Figure 5.5 *Micellar electrokinetic chromatograms of FMOC-amino acids. Conditions; A. buffer, 50 mM phosphate (pH = 7.5), 50 mM sodium dodecylsulfate, 12 mM γ-CD, 15% (v/v) 2-propanol; current, 11 μA; column, 67 cm × 25 μm ID (45 cm to detector). 25 kV; temperature, 25 °C; UV detection at 256 nm. B. 12 mM β-CD, other conditions as in A (see reference for chromatographic conditions)* (Reprinted with permission from ref. 31)

being more effective than γ-CD. The efficiency of enantiomeric separation was considerably enhanced by the addition of 2-propanol to the running buffer.

Native and substituted β-CDs have also been employed for the chiral separation of 6-aminoquinolyl-N-hydroxysuccinimidyl carbamate (AQC) derivatised amino acids.[32] The background electrolyte was 10 mM aqueous 1,3-bis-[tris(hydroxy-methyl)methylamino]propane (BTP) (pH 7.0) containing 5 mM of CDs, including β-CD, β-CD polymer crosslinked with 1-chloro-2,3-epoxypropane, HP-β-CD, dimethyl-β-CD (DM-β-CD) trimethyl-β-CD (TM-β-CD) and carboxymethyl-β-CD polymer crosslinked with 1-chloro-2,3-epoxypropane (CM-β-CD polymer). The same methods have also been employed for the determination of the complexation constants of AQC-amino acids with various substituted β-CD derivatives.[33] The effective selectivity coefficient, α^{eff}, can be calculated by

$$\alpha^{eff} = (\mu^{free} + \mu_2{}^{cplx} K_2[S])/(\mu^{free} + \mu_1{}^{cplx} K_1[S])(1 + K_1[S])/(1 + K_2[S]) \quad (5.2)$$

where μ^{free} is the effective electrophoretic mobility of enantiomers in the background electrolyte in the absence of CD, $\mu_1{}^{cplx}$ and $\mu_2{}^{cplx}$ are the electrophoretic mobilities of enantiomers in the presence of CD, K_1 and K_2 are the complexation constants of enantiomers, and [S] is the concentration of CD in the background electrolyte. The type of CD derivative exerted a considerable influence on the enantioselective separation of AQC-amino acids.

The effect of HP-β-CD concentration on the electrophoretic mobility of dansylated amino acids has been studied in detail.[34] It was established that the differences between the electrophoreretic mobilities of enantiomers increased with increasing concentration of HP-β-CD in the lower concentration range, reached a maximum then decreased at higher HP-β-CD concentrations. Phenylalanine enantiomers showed irregular behaviour at pH 2.75, the differences reached a plateau and did not decrease at higher concentations of HP-β-CD.

The enantiomeric separation of phenylthiohydantoin-amino acids (PTH) has also been performed by chiral CE.[35] Separations were carried out on an uncoated capillary. Amino acid derivatives were injected at the cathode employing a gravity method. The efficiencies of the seven chiral selectors. β-escin, digitonin, TM-β-CD, β-CD, α-CD, saponin, and glycyrrhizin ammonical hydrate were compared using running buffers of different compositions. The results showed that a single chiral selector is not suitable for the separation of every enantiomeric pair; PTH-amino acids can be separated only by using four different chiral selectors. However, it was found that running buffers containing more than one chiral selector were able to separate the majority of enantiomeric pairs (Figure 5.6). The method was proposed for protein sequence analysis and for the analysis of free amino acids in biological fluids after improving the enantiomeric separation capacity and control of racemisation.

The separation capacity of γ-cyclodextrin sulfobutyl ether (γ-CD-SBE) and native γ-CD was compared using dansylated amino acids and other racemic compounds as model solutes.[36] An untreated fused-silica capillary was employed for the enantiomeric separations). Dansyl-$tert$-Leu was separated in 30 mM phosphate buffer (pH 7) containing either 10 mM γ-CD-SBE or 20 mM γ-CD and

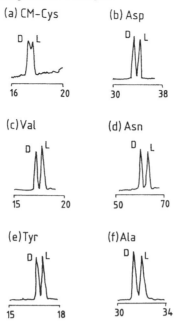

Figure 5.6 *Optical resolution of typical PTH-DL-amino acids using a mixed electrolyte (10 mM formic acid–17.5 mM TM-β-CD–12.5 mM digitonin–12.5 mM β-escin–50 mM SDS). (CM = carboxymethyl) on uncoated capillary 50 cm × 30 cm 50 μm internal diameter. Separation voltage was −15 kV* (Reprinted with permission from ref. 35)

50 mM SDS. All other dansylated amino acids were separated in 30 mM phosphate buffer (pH 7) methanol (9:1, v/v) containing either 5 mM γ-CD-SBE or 20 mM γ-CD and 50 mM SDS. The optimal concentration of γ-CD-SBE is generally lower than that of γ-CD, especially in the case of cationic solutes; its application as chiral selector in CE is proposed.

A special CE method has been developed for the quantitative determination of low quantities of L-Trp in D-Trp.[37] The relative standard deviations of the validation parameters were 0.73% (migration time), 0.60% (relative migration time), 8.16% (peak area) and 0.80% (peak area ratio). The relationship between the concentration of D-Trp and the detector response (peak area) was linear in the concentration range of 0.4–2.5 mg mL^{-1}, the correlation coefficient being 0.9992. The precision of response factors was 1.45% and the method was successfully applied on different CE instruments (Figure 5.7).

Not only amino acids but also peptides were separated by CE using CDs as buffer additives. Thus, neutral and charged CDs have been employed for the separation of stereoisomers of aspartyl di- and tri-peptides.[38] Fused-silica and polyacrylamide-coated capillaries were used in the experiments. The running buffer consisted of 50 mM sodium phosphate buffer at various pH containing

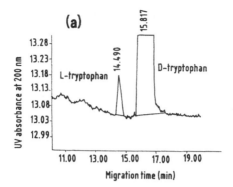

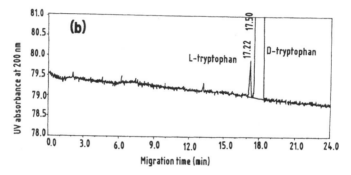

Figure 5.7 *Separation of 0.1% L-tryptophan in D-tryptophan. (a) Beckman instrument; (b) ABI instrument. Separation conditions; 1.5 mg mL^{-1} tryptophan solution in electrolyte*
(Reprinted with permission from ref. 37)

β-CD, CM-β-CD and β-CD-SBE. It has been established that Asp di- and tri-peptides can be easily separated by CM-β-CD and β-CD-SBE with different degrees of substitution in both fused-silica and polyacrylamide-coated capillaries. A similar method was employed for the enantiomeric separation of tripeptide enantiomers using CM-β-CD.[39] It was found that the change of buffer pH may result in the reversal of migration order for one enantiomeric pair of tripeptides (Figure 5.8). The reversal of migration order was tentatively explained by the supposition that the stability of the inclusion complexes changes differently with pH resulting in modified migration velocities. Sulfated CDs have also been applied for the separation of aspartyl di- and tri-peptides by CE.[40] Experiments were performed in an uncoated fused silica at -20 kV and 25 °C. Highly sulfated (HS) α-, β- and γ-CDs (degree of substitution 12) and another sulfated β-CD (degree of substitution 4) were added separately to the running buffer (triethylammonium phosphate, pH 3). The results showed that each sulfated CD derivative was able to separate the peptide enantiomers. However, the best separation of

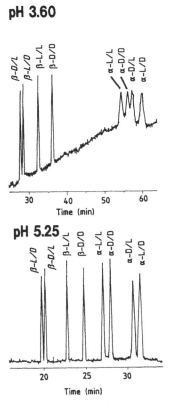

Figure 5.8 *Separation of the stereoisomers of the isomeric tripeptides Gly-α/β-D/L-Asp-D/L-PheNH₂ in polyacrylamide-coated capillaries at pH 3.60 and 5.25. Conditions: 47/40 cm capillary, 50 mM sodium phosphate buffer, 60 mg mL⁻¹ CM-β-CD, −20 kV on 50 μm ID acrylamine-coated capillaries, detection 215 nm (Reprinted with permission from ref. 39)*

tripeptides was obtained by the application of HS-γ-CD. *N*-(substituted)-glycine-peptoid (NSG) combinatorial mixtures have also been separated by CE using methyl-β-CD as buffer additive and the results were compared with those obtained by HPLC.[41] CE separations of the sample CHIR 4582 were performed in fused-silica capillaries at 15 kV (Figure 5.9). HPLC was carried out on a C_{18} column. Solvent A was 5% acetonitrile in 0.2% heptanesulfonic acid–0.08% trifluoroacetic acid and solvent B was 75% acetonitrile in 0.2% heptanesulfonic acid–0.08% trifluoroacetic acid. Initial mobile-phase composition was 0% B, to 100% B in 35 min at a flow-rate of 50 μL min⁻¹. Solutes were well separated, indicating that CE and HPLC can be equally employed for the analysis of these types of peptide mixtures.

Dansylated amino acids and other racemic compounds have been separated with sulfobutyl CD derivatives and the degree of substitution on the efficiency of

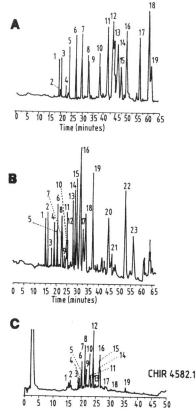

Figure 5.9 *CE separation of the linear trimeric NSG-peptoid mixture; CHIR 4582. Running buffer conditions were (A) 250 mM sodium phosphate buffer, pH 2.0, with 25 mM HSA and 40 mg mL⁻¹ methyl-β-CD; (B) 200 mM sodium phosphate buffer, pH 2.0, with 75 mM HSA and 30 mg mL⁻¹ methyl-β-CD; (C) Reversed-phase HPLC chromatogram. ISCO Model 3850 capillary electropherograph on fused-silica capillaries, separation voltage 15 kV*
(Reprinted with permission from ref. 41)

separation has been studied in detail.[42] It was established that γ-CD-SBE is an efficient chiral selector, and the degree of substitution has a considerable influence on the enantiomeric selectivity. Enantiomeric separation of dansyl-amino acids has also been obtained in a non-aqueous media (N-methylformamide) using β-CD as chiral selector.[43] Sulfobutyl ether-β-CD has also been employed for the chiral separation of dansyl-amino acids and pharmaceuticals, and the marked effect of pH on the enantioselectivity has been demonstrated.[44] A similar effect of pH was found in the CE analysis of dansylated amino acids and pharmaceuticals using with native CDs as chiral selectors,[45] and in the analysis of Trp derivatives in the presence of a soluble β-CD polymer.[46] The effect of unsubstituted, methyl-, ethyl- and 1,3-dimethyl-ureas on the enantioselectivity of native, di- and tri-methyl CDs has been studied in detail. Dansylated amino acids were employed as model compounds. The results indicated that ureas enhance the chiral selectivity of native CDs but decrease that of methylated CD derivatives.[47] A similar method has been employed for the elucidation of the effect of methylation on the enantioselectivity of γ-CD derivatives using dansylated amnio acids as model compounds. It was established that the position of the methyl group exerts a considerable effect on the chiral resolution.[48] Selectively methylated β-CD derivatives showed similar behaviour.[49] Theoretical optimisation procedures for CD mediated CE[50] and a systemic approach for the treatment of chiral separation[51] have been previously published. Not only dansylation but also other precolumn derivatisation methods have been applied for racemic amino acids. Thus, the separation of 3,5-dibenzamido Phe enantiomers has been obtained by using β-CD[52] as a chiral selector. The effect of the derivatisation with 9-fluoromethyl chloroformate using β-CD,[53] and with 2-(9-anthryl)ethyl chloroformate using β- and γ-CD,[54] on the chiral resolution has been demonstrated. Naphthalenesulfonic acids[55] and N-(3,5-dinitrobenzoyl)amino acids[56] have also been separated with methylated CDs and 1-(1-naphthyl)ethylcarbamoylated β-CDs, respectively. The resolution of various amino acid derivatives (dinitrobenzoyl, dinitrophenyl, dimethylaminonaphthylsulfonyl, carboxybenzyl, 9-fluoromethoxycarbonyl and 6-aminoquinolyl-N-hydroxysuccinimidylcarbamoyl) using modified CDs has been studied in detail. The results proved that derivatives with nitro- or dimethylamino-groups were better separated. Methylated and hydroxypropylated CDs showed higher enantioselectivity.[57] Biologically active penta- and nona-peptides have been resolved at acidic pH adding 10 mM β-CD to the buffer.[58] The enantiomers of the tetrapeptide Tyr-D-Arg-Phe-Phe-HN$_2$ have been resolved in the presence of 10 mM DIMEB in the running buffer.[59] Direct separation of 16 derivatised di- and tri-peptides was achieved by adding CD, vancomycin or teicoplanin to the running buffer. The enantioselectivity depended on the chiral selector analyte pair.[60] The use of packed capillary electrochromatography for the enantiomeric separation of dansyl- and dinitrophenylamino acids on a β-CD bonded stationary phase has also been reported.[61]

Separation of Positional and Optical Isomers of Pharmaceuticals

Because of the well-known diverse biological activity of drug enantiomers much effort has been devoted to the development and application of various CE methods for the chiral separation of racemic pharmaceuticals. A considerable number of drugs showing widely different pharmaceutical activity have been investigated by CE techniques.

The enantiomeric separation of non-steroidal anti-inflammatory drugs (NSAIDs) has been extensively studied. Thus, the chiral separation of ibuprofen, 1-phenylethanol, flurbiprofen, cyclofen and etodolac has been well illustrated (Figure 5.10).[62] The results indicated that increasing film thickness resulted in higher retention factors but the theoretical plate number and the separation factor decreased at higher film thickness (Table 5.1). Buffers containing more than one

A.   **B.**

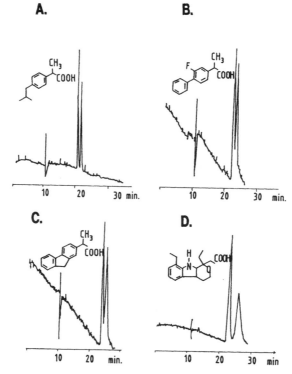

C.   **D.**

Figure 5.10 *Enantiomeric separation of racemic ibuprofen (A), flurbiprofen (B), ciclopro-*
fen (C), and etodolac (D). Conditions; A, B and C, Chirasil-Dex based on
permethylated β-cyclodextrin (d_f = 0.4 μm); pH 7.0; applied voltage; 30 kV.
D, Chirasil-Dex based on permethylated γ-cyclodextrin. Other conditions are
the same as for A, B, and C
(Reprinted with permission from ref. 62)

Table 5.1 *Effect of film thickness (d_f) on the separation parameters*

Analyte	d_f (μm)	k_2	α	N
1-Phenylethanol	0.2	0.21	1.17	30500
	0.4	0.32	1.19	24016
	0.8	1.10	1.12	9507
Flurbiprofen	0.2	1.03	1.04	31162
	0.4	1.36	1.05	12492
Cicloprofen	0.2	1.14	1.06	19828
	0.4	1.43	1.08	9939

(Reprinted with permission from ref. 62)

CD derivative have also been employed for the chiral separation of NSAIDS.[63] The mobility of NSAIDs increased with increasing concentration of β-CD-NH$_2$ in the running buffer at the lower concentration range (1–5 mM) but then did not change at 5–20 mM concentrations. Better enantiomeric separations were achieved by using a dual CD system of 20 mM β-CD-NH$_2$/TM-β-CD. Another study showed that TM-β-CD was a better chiral selector for 2-arylpropionic acid non-steroidal anti-inflammatory drugs than DM-β-CD and 6^A-methylamino-β-CD.[64] The thermodynamic parameters of the interaction of ibuprofen enantiomers with β-CD have also been measured.[65] It was established that the electrophoretic mobilities have a temperature coefficient of *ca.* 2% while the pK of ibuprofen was practically temperature independent (4.37 and 4.40 at 25 and 50 °C, respectively). The efficacy of chiral separation decreased markedly at higher temperatures (Figure 5.11). CE has also been employed for the analysis of ketoprofen in serum and the results were compared with those obtained by HPLC (Figure 5.12).[66] RSD values were 2.7% and 6.0% for HPLC and CE, respectively. The CE method was cheaper than HPLC, but the sensitivity and reproducibility were lower. CE was proposed as a complementary method for the determination of interferences and impurities.

The chiral separation capacities of three enantiomeric selectors (TM-β-CD, heptamethylamino-β-CD and vancomycin) were compared using racemic NSAIDs (Figure 5.13).[67] The electropherograms differ; however, each chiral selector was suitable for the enantiomeric separation of both indoprofen and flurbiprofen. The

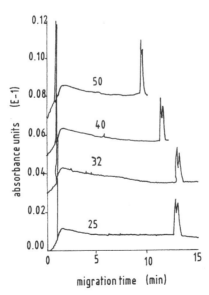

Figure 5.11 *Separation of a racemic mixture of ibuprofen in 0.01 M sodium/acetate, pH 4.47 with 2.5 mM β-CD in a coated capillary at different temperatures* (Reprinted with permission from ref. 65)

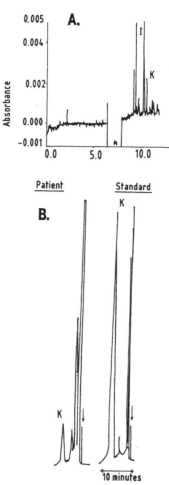

Figure 5.12 *Electropherogram (A) and chromatogram (B) of serum from a patient taking 25 mg of ketoprofen (serum level 1.6 mg L⁻¹). CE separation was performed in a untreated capillary at 12 kV and 30 °C. Buffer consisted of 250 mM boric acid, pH adjusted to 8.9 with 2 M sodium hydroxide; ACN (10 mL L⁻¹) and β-CD (1 g L⁻¹) were added to the buffer. HPLC experiments were carried out on a Econosphere CN column. Mobile phase was 1000 mL of water, 40 mL of ACN and 200 μL of phosphoric acid. Detection wavelength was 254 nm. A: injection 99 s; B: arrows indicate injection. I = internal standard, K = ketoprofen*
(Reprinted with permission from ref. 66)

RSD values of the determination of the migration time were 0.37–1.01% showing the good reproducibility of the CE system. The column efficiency was the lowest for heptamethylamino-β-CD (55–84) followed by TM-β-CD (68–113) and vancomycin (249–269). Because of the better separation capacity and lower

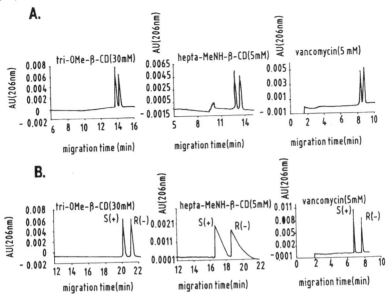

Figure 5.13 *Electropherograms of the enantiomeric separation of indoprofen (panel A) and flurbiprofen (panel B) using 75 mM buffer (pH 5) and different chiral selectors. Capillary, 35 (31.5) cm × 50 μm ID (coated); −20 kV, 32.8– 33.6 μA; injection at 10 psi; indoprofen and flurbiprofen (5.10^{-5} M)* (Reprinted with permission from ref. 67)

detection limit the use of vancomycin was proposed for the chiral separation of NSAIDs. Chiral separation of naproxen and fenoprofen has also been achieved by using HP-β-CD as a chiral disciminator.[68] The new cationic β-CD derivative heptakis-(6-methoxyethylamine)-β-CD (β-CD-OMe) has also been successfully applied for the enantiomeric separation of NSAIDs and phenoxypropionic acid herbicides (Figure 5.14).[69] The migration time of NSAIDs increased with increasing concentration of β-CD-OMe in the running buffer (50 mM NaH$_2$PO$_4$, pH 5) between 1 and 5 mM. The shortest migration time was observed at pH 5 and it increased at both higher and lower pH.

Fenoprofen, together with other acidic and basic enantiomers, was separated by using methylamino-β-CD as chiral selector (Table 5.2).[70] Resolution increased with increased concentration of MeNH-β-CD in the buffer or reached a maximum and then decreased at higher CD concentrations. The pH of the buffer between 5–7 exerted a negligible effect on the resolution. Terbutaline, chlorpheniramine, isoproterenol, ketamine and propranolol were separated in an untreated capillary in 100 mM H$_3$PO$_4$/TMA containing 1–7.5 mM MeNH-β-CD. The applied voltage was 18 kV; however, the migration time was fairly long. Enantiomers of fenoprofen and ibuprofen have also been separated by β-CD and the effect of various separation parameters on the resolution has been studied in detail.[71–73] Chiral separation of six phenoxypropionic acid herbicides [PPAHs: 2-(3-

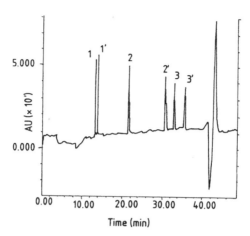

Figure 5.14 *Optimised Enantiomeric separation of a standard mixture of six (+) NSAID enantiomers. The background electrolyte contains 50 mM NaH₂PO₄, pH 6; 3 mM β-CD-OMe; applied voltage was −30 kV, −25 μA; pressure injection 85 kPa s; sample concentration 0.1 mg mL⁻¹ in methanol-water (1:1, v/v); (1,1′) = ketoprofen; (2,2′) = fetoprotein; (3,3′) = flurbiprofen* (Reprinted with permission from ref. 69)

Table 5.2 *Comparison of enantiomeric resolution of acidic compounds using 5 mM of MeNH-β-CD and (MeNH)7-β-CD. Apparatus: Bio-focus 3000; capillary: coated 35(30.5) cm × 50 m I.D.; background electrolyte 75 mM and 5 mM of CD; applied voltage: 15 kV. Injection: pressure 10 psi s*

Compounds	CD type MeNH-β-CD 5 mM	(MeNH)7-β-CD 5 mM
	Resolution (R)	
Phenyllactic acid	3.27	3.56
3-Isomer of tiaprofenic acid	no	0.6
Tiaprofenic acid	1.50	1.76
Fenoprofen	0.50	1.37
Warfarin	0.93	0.40
Acenocoumarol	1.40	no

(Reprinted with permission from ref. 70)

chlorophenoxy)propionic acid; 2-(2-chlorophenoxy)propionic acid; 2-(4-chloro-phenoxy)propionic acid; 2-(2,4-dichlorophenoxy)propionic acid; 2-(2,4,5-tri-chlorophenoxy)propionic acid; 2-(2-phenoxy)propionic acid] was achieved under similar conditions. It was concluded that this new cationic β-CD derivative is suitable for chiral separation over a wide range of pH and concentrations. The results further indicated that cationic α- and γ-CD derivatives can also be employed for the enantiomeric separation of small and large, charged and uncharged guest molecules.

Non-aqueous capillary electrophoresis (NACE) has also found application for the enantiomeric separation of NSAIDs and other racemic compounds using heptakis-(2,3-diacetyl-6-sulfato)-β-CD (HDAS-β-CD) as a chiral resolving agent.[74] The use of NACE was required because neutral and weak electrolyte enantiomers are sometimes poorly soluble in aqueous running buffers. Separations were performed in an untreated fused-silica capillary. The chemical structures of the analytes are found in Figure 5.15. The background electrolyte was a mixture of dichloroacetic acid and triethylamine dissolved in methanol. It was found that NACE using HDAS-β-CD produces baseline, rapid and rugged separation of weak base enantiomers in acidic methanol background buffer when the concentration of β-CD derivative is between 20 and 50 mM.

Chlorthalidone enantiomers have also been separated by capillary electrochromatography adding HP-β-CD[75] or mono-(6-amino-6-deoxy)-β-CD as chiral selectors.[76]

CDs have also been used for the chiral separation of racemic anaesthetics by CE. Thus, mixtures of anionic and neutral β-CDs have been employed for the enantiomeric separation of barbiturates and other pharmaceuticals (Figure 5.16).[77] It was stated that the neutral CDs were unsuitable for the enantiomeric separation of drugs at pH 3; however, negatively charged β-CD derivatives showed marked chiral selectivity. Alkaline pH buffers reduced the efficiency of separation. The resolution depended highly on the CD concentration. Dual systems showed better chiral selectivity than buffers containing only one CD, the pH again influencing the efficiency of separation.

Combinations of CDs have also been employed for the chiral separation of acidic drugs, such as sulindac, fenoprofen, ketoprofen, warfarin and hexobarbital, using the same conditions.[78] Addition of neutral β-CDs to the anionic β-CD derivative considerably improved the separation, the effect depending on the type of neutral CD (Table 5.3). It was assumed from the data that the anionic CD provided the electromobility while uncharged CDs accounted for the enantioselectivity of separation.

The effect of degree of substitution and substitution pattern of methylated β-CD derivatives on the enantioselectivity has been studied in detail.[79] Five different DM-β-CDs and two different randomly methylated β-CDs (RAM-β-CD) were separately added to the running buffer at a concentration of 15 mM and compared with HPLC separations performed on a C_4 column mobile phase using 10 mM phosphoric acid containing 1 mM methylated CDs. The results indicated that methylated β-CDs influence the migration and retention times of analytes differently and their effect on the efficiency of enantiomeric separation is also different. It was further established that HPLC was less sensitive to the degree of substitution and the substitution pattern than CE.

A positively charged β-CD derivative (2-hydroxypropyltrimethylammonium salt, TMA-β-CD) has also been employed for the enantiomeric separation of hexobarbital and other pharmaceutics.[80] TMA-β-CD was added to the background

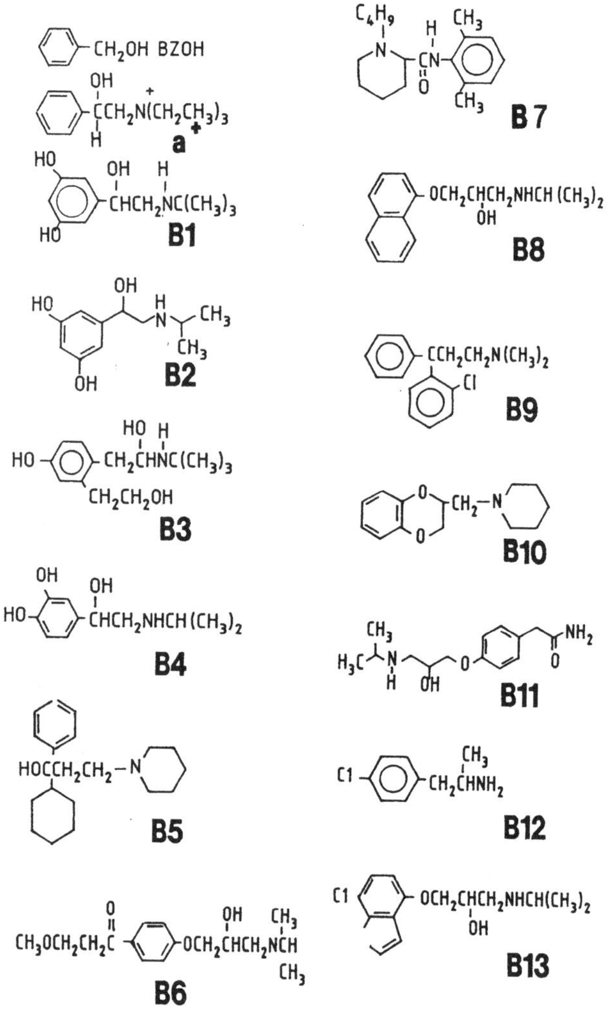

Figure 5.15 *Chemical structure of analytes for nonaqueous capillary electrophoresis*
(Reprinted with permission from ref. 74)

electrolytes in different concentrations. Electropherograms using TMA-β-CD proved that this β-CD derivative contains a minimum of four different fractions (Figure 5.17). The electroosmotic flow (EOF) markedly depended on the concentration of TMA-β-CD in the running buffer: a cathodic EOF was observed at low concentrations and an anodic one at higher concentration of TMA-β-CD. It was established that peak efficiency increased with increasing concentration of TMA-β-CD but the enantioselectivity and migration time decreased. It was also observed that the enantiomeric migration order of neutral and anionic solutes may be reversed at different concentrations of TMA-β-CD. Because of its good

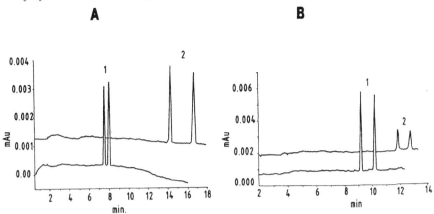

Figure 5.16 *Enantioseparation of pharmaceuticals in dual CD systems. (A) Pentobarbital. Buffers; (1) 100 mM phophoric acid–triethanolamine (pH 3) containing β-CD-SBE (5 mM) and TM-β-CD (30 mM); (2) 100 mM phosphoric acid–triethanolamine (pH 5) containing CM-β-CD (10 mM). (B) Mephenytoin (1), chlorthalidone (2). Buffer; 100 mM phosphoric acid–triethanolamine (pH 3) containing CM-β-CD (10 mM) and TM-β-CD (10 mM)*
(Reprinted with permission from ref. 77)

enantioselectivity the application of the positively charged TMA-β-CD in chiral CE is highly recommended.

The enantiomers of the local anesthetic prilocaine were determined in human serum using CDs as chiral selectors.[81] Prilocaine was extracted from human serum and 40 μL solution of procainamide internal standard. Running buffer was 100 mM sodium dihydrogen phosphate, pH 2.5 containing 15 mM CDs and 0.03 mM hexadecyltrimethylammonium bromide (HTAB). Surprisingly, only DM-β-CD was suitable for the enantiomeric separation of prilocaine: native α-, β- and γ-CD, TM-β-CD, HP-β-CD, HP-α-CD, HP-γ-CD, hydroxyethyl-β-CD (HE-β-CD), CM-β-CD and amino-β-CD were ineffective (Figure 5.18). The validation parameters of the method were very good. Recoveries were $85.2 \pm 3.9\%$ and $84.9 \pm 4.8\%$ for R-(−) and S-(+)-prilocaine, respectively. The reproducibility of the migration time for each analyte was lower than 0.5%. The limits of detection and limits of quantitation were 38 and 45 ng mL^{-1}, respectively. Because of the good validation parameters the CE method was proposed as a suitable alternative to HPLC for the enantioselective analysis of prilocaine in human serum.

Cyclodextrins have also been used as chiral additives for the enantiomeric separation of epimeric *N*-oxides on morphinane alkaloids.[82] Pyridoxol was used as an internal standard. The results clearly indicated that the separation capacity of CDs is different for different stereoisomers to be separated. The best results were achieved by using γ-CD and DM-β-CD. The combination of chiral and achiral ion-pairing agents and cyclodextrins has been employed for the enantiomeric separation of the anticholinergic drug cyclopentolate and other racemic

Table 5.3 *Influence of the addition of a neutral β-cyclodextrin in a buffer containing β-CD-SBE on enantioresolution*

	Enantiomeric resolution (R_s)					
	β-CD-SBE	β-CD-SBE/β-CD	β-CD-SBE/M-β-CD	β-CD-SBE/DM-β-CD	β-CD-SBE/TM-β-CD	β-CD-SBE/HP-β-CD
Sulindac	1.4	2.3	3.5	3.8	1.6	<0.7
Fenoprofen	<0.7	1.0	1.5	2.8	2.8	<0.7
Ketoprofen	1.1	1.5	2.1	4.6	2.1	1.3
Warfarin	2.2	5.1	9.2	9.3	4.3	2.7
Hexobarbital	1.7	0.8	1.2	3.0	1.9	<0.7

Buffer; 5 mM β-CD-SBE in 100 mM phosphoric acid adjusted to pH 3.0 with triethanolamine containing no additional cyclodextrin, or β-CD, M-β-CD, DM-β-CD, TM-β-CD and HP-β-CD (10 mM).
(Reprinted with permission from ref. 78)

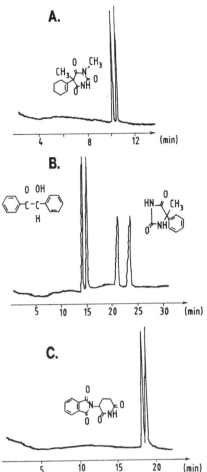

Figure 5.17 *Enantioseparation of ±-hexobarbital (A, 10 mg mL⁻¹ TMA-β-CD), (±)-ben-zoin and (±)-5-methyl-5-phenylhydantoin (B, 50 mg mL⁻¹ TMA-β-CD), and (±)-thalidomide (C, 20 mg mL⁻¹ TMA-β-CD). Conditions; 50 mM phosphate buffer, pH 3.0, +400 V cm⁻¹, detection, 210 nm, polyacrylamide-coated capillary (50/43 cm)*
(Reprinted with permission from ref. 80)

compounds.[83] Separations were performed on untreated and neutrally coated capillaries. CE conditions varied according to the analytes to be separated. The various combinations of ion-pairing reagent (IPR) and cyclodextrin resulted in different enantioselectivity (Figure 5.19). It was assumed that the simultaneous application of CDs and ion-pair reagents offers a promising possibility for the enantiomeric separation of a wide variety of racemic compounds.

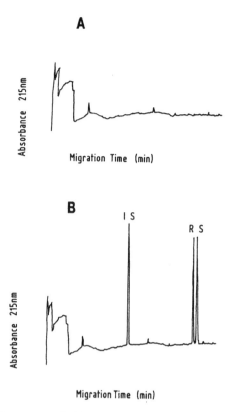

Figure 5.18 *Electropherograms of (A) serum blank and (B) serum spiked with R-(−)-prilocaine (400 ng mL⁻¹, 16.0 min), S-(+)-prilocaine (400 ng mL⁻¹, 16.4 min) and procainamide (internal standard, 400 ng mL⁻¹, 8.9 min). Electrophoretic conditions; uncoated fused-silica capillary (50 μm ID, 72 cm total length, 50 cm effective length), the background electrolyte consisted of an aqueous solution of 100 mM sodium dihydrogen phosphate, pH 2.5 (adjusted with 100 mM phosphoric acid), containing 15 mM heptakis (2,6-di-O-methyl)-β-cyclodextrin (DM-β-CD) and 0.03 mM HTAB; voltage, 25 kV; temperature, 30 °C; vacuum injection, (5' Hg), 24 s, detection was at cathodic end* (Reprinted with permission from ref. 81)

The cationic CD derivative (3-hydroxy-3-trimethylammoniopropyl-β-CD, TMA-β-CD) has also been employed for the enantiomeric separation of pharmaceuticals using the same CE conditions.[84] It was established that TMA-β-CD is also suitable for the chiral separation of basic and acidic analytes such as cyclodrine, phendimetrazine, cyclopentolate, pholedrine, norpseudoephedrine, terbutaline, *p*-bromomandelic acid, 2-phenylbutyric acid, mandelic acid, 2-phenylpropionic acid, 2-methylphenylacetic acid and tropic acid. TMA-β-CD causes the reversal of EOF and migrates in the opposite direction to EOF. The migration order of enantiomers can be reversed by changing the pH value.[85] Another set of racemic anticholinergic drugs was separated by CE using native and substituted

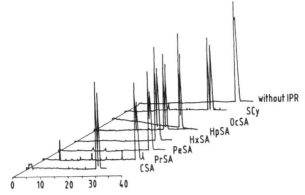

Figure 5.19 *Electropherograms of cyclodrine with different buffer additives 100 mM phosphate buffer (PB), 1.8% β-CD, 40 mM alkanoic acid, pH 2; 214 nm, 35 °C; 12 kV; uncoated capillary 57/50 cm × 75 μm. IPR = ion pairing reagent; SCy = sodium cyclamate; OcSA = octaniesulfonic acid; HpSA = heptanesulfonic acid; HxSA = hexanesulfonic acid; PeSA = pentanesulfonic acid; PrSA = propanesulfonic acid; CSA = camphorsulfonic acid* (Reprinted with permission from ref. 83)

CDs (Figure 5.20).[86] It was found that baseline separation of anisodamine enantiomers was achieved but the enantiomers of atropine were not separated using CDs as chiral selectors. The same method was successfully applied for the chiral separation of methoxamine and lobeline.[87]

The enantiomeric separation of atropine (hyoscyamine) has also been achieved by using CDs as chiral selectors in CE.[88] Atropine was extracted from *Scopolia* extract and *Scopolia rhizome* by mixing 10% ammonia solution. The effect of temperature on the enantiomeric separation was studied by thermostating the capillary between 15 and 50 °C. The buffer consisted of 100 mM sodium phosphate solution. Resolutions and migration times of hyoscyamine enantiomers were measured in the presence of various CDs (Table 5.4). It was established that low pH and higher buffer concentration up to 100 mM improved resolution, while an increase of temperature exerted an opposite effect on the chiral separation. Coefficients of correlation were high in the concentration range 10–200 μg mL^{-1} ($r \geqslant 0.99$) for both enantiomers of atropine and for the internal standard scopolamine. The reproducibilities of migration time were 0.88–0.96% and those of peak area 1.03–6.35%. It was stated that the method complies with the requirements of *The Japanese Pharmacopoeia* and can be employed for the separation and quantitative determination of hyoscyamine enantiomers in pharmaceutical preparations.

The enantiomeric separation of the antihistaminic drug dimethindene (DIM has also been achieved by using various CDs as chiral selectors (Figure 5.21).[89] The concentration of CDs providing adequate chiral separation depended considerably on the type of substitution. A reversal of migration order also occurred in some

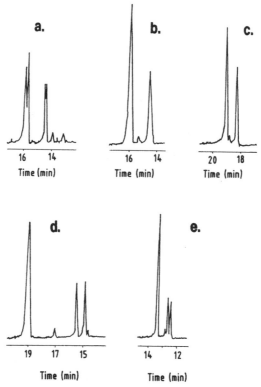

Figure 5.20 *Electropherograms of the chiral separation of racemic glycopyrronium. Background electrolyte; 100 mM Tris-phosphoric acid buffer (pH 2.3) containing 12 mM α-CD (a), 12 mM β-CD (b), 12 mM DM-β-CD (c), 12 mM HP-β-CD (d), and 12 mM γ-CD (e), separation tube, 62 cm (41 cm to detector) × 50 μm I.D.; running voltage 22 kV, detection, 210 nm (0.005 AUFS), temperature, ambient (about 15 °C)*
(Reprinted with permission from ref. 86)

instances. NMR studies of DIM CM-β-CD mixtures suggested the possibility of the multimodal complex formation with complexes of different stoichiometry. It was assumed that this phenomenon may account for the high chiral separation capacity of CM-β-CD and the reversal of retention order. Electrospray mass spectrometry (ESI-MS) combined with CE has been employed for the chiral separation of DIM and other racemic compounds.[90] CE-ESI-MS separations (Figure 5.22) were carried out in untreated fused-silica and polyacrylamide-coated capillary for basic and acidic analytes, respectively. The high sensitivity of ESI-MS detection was demonstrated in the enantiomeric separation of dimethindene, chlorpheniramine (0.2 mg mL^{-1} CM-β-CD for both), and tropic acid (8 mg mL^{-1} TMA-β-CD). Mianserin enantiomers have been previously resolved using native, sulfobutyl and sulfoethyl β-CD.[91]

Table 5.4 *Resolution and migration time of D,L-hyoscyamine*

Cyclodextrin	Concentration used (mM)	Migration time (min)		Resolution
		L-HYO[a]	D-HYO[b]	
α-CD	30	16.6	–	0.00
β-CD	15	20.8	–	0.00
γ-CD	30	21.4	21.6	0.55
M-β-CD	30	30.9	31.1	0.55
DM-β-CD	30	36.5	–	0.00
TM-β-CD	30	15.7	15.9	1.53
HP-β-CD	30	33.0	–	0.00

[a] L-HYO, L-hyoscyamine; [b] D-HYO, D-hyoscyamine.
(Reprinted with permission from ref. 88)

Chiral CE and HPLC methods were developed for the study of the *in vitro* metabolism of the racemic antiasthmatic/antiallergic drug flezelastine.[92] Human, rat, bovine and porcine liver microsomes were employed for biotransformation studies of flezelastine. It was found that liposome preparations metabolise flezelastine in different ways. Chiral HPLC proved that human and rat liposomes metabolise the enantiomers at similar rates while bovine and porcine liposomes showed stereoselective metabolism. It was found that the enantiomers were better separated under CE than under HPLC conditions (Figure 5.23) indicating that CE is more suitable for the enantiomeric separation of racemic flezelastine than chiral HPLC.

Water-soluble melatonergic drugs have also been separated with CDs added to the background electrolyte.[93] Enantiomeric separation was only achieved in the presence of CDs. Chiral separation was followed by the reversal of migration order. The optimal concentration for β-CD and DM-β-CD was 15–28 mM. Binding constants for β-CD determined by a non-linear curve fitting procedure were 60.8 M^{-1} and 95.1 M^{-1} for BMS-191435(R) and BMS-191435(S), respectively. The binding constants for DM-β-CD were 67.4 M^{-1} and 85.5 M^{-1}. The energy difference between the two enantiomers was *ca.* 0.3 kcal mol^{-1}. Resolution depended highly on the pH, the optimum being 2.58 for DM-β-CD and 4.10 for β-CD. It was assumed that the formation of hydrogen bonds between the secondary aliphatic amines of drugs and the CD molecule plays a decisive role in the chiral separation. Computer modelling indicated that the hydrophobic substructures of enantiomers are included in the CD cavity and the hydrophilic part forms hydrogen bond with the polar hydroxyl groups at the outer sphere of the CD.

Nuclear magnetic resonance was applied for the screening of CDs for the design of chiral separation of the racemic anti-schistosomiasis drug oxamniquiune.[94] The oxamniquine CD 1:1 complexes were studied by 1D NMR. The

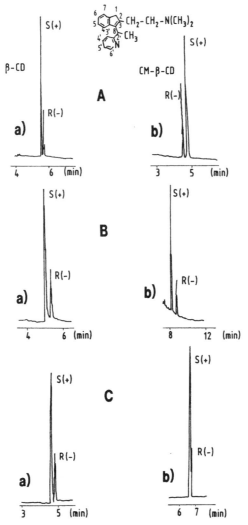

Figure 5.21 *Electropherograms of the mixture of DIM enantiomers [R(−)/S(+) = 1/3] in the presence of (Aa) 15 mg mL⁻¹ β-CD; (Ab) 1 mg mL⁻¹ CM-β-CD; (Ba) 15 mg mL⁻¹ 6-deoxy-6-monocarboxy-β-CD; (Bb) 5 mg mL⁻¹ β-CD-6-mono-phosphate; (Ca) 15 mg mL⁻¹ β-CD in the absence of 5 M urea; (Cb) 15 mg mL⁻¹ β-CD in the presence of 5 M urea. Buffer: 50 mM potassium phosphate at pH 3.0*
(Reprinted with permission from ref. 89)

diagnostic aromatic singlets indicating interaction with each CD is, however, according to the shift non-equivalence for the two aromatic singlets, only showing anionic CD interaction. It was concluded from the NMR data that anionic CDs are probably better chiral selectors than neutral ones. CE experiments were

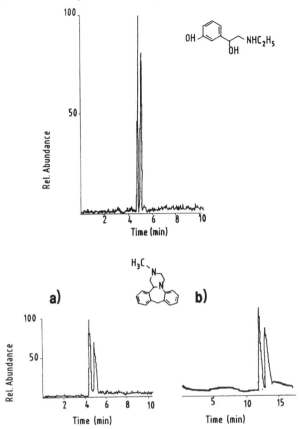

Figure 5.22 *Upper part: CE-ESI-MS electrophoregram of (±)-etilefrine plotted in the single ion track (m/z = 181.7–182.7) mode with 3 mg mL⁻¹ CM-β-CD as a chiral selector (pH 4.3). Lower part: CE-ESI-MS electrophoregram of (±)-mianserine plotted in the single ion track (m/z = 264.9–265.9) mode (a) and CE-UV electropherogram (b) with 0.2 mg mL⁻¹ CM-β-CD as a chiral selector* (Reprinted with permission from ref. 90)

performed in an uncoated fused-silica capillary. The results indicated that not only the type of substituent but also the degree of substitution exerted a considerable effect on the efficiency of chiral separation. Oxamniquine enantiomers have been previously separated with β- and hydroxypropyl-β-CD as chiral selectors.[95]

CDs in combination with Zn(II)–phenylalanine (Phe) have been employed for the chiral separation of new antibacterial quinolone drugs (NQs).[96] CE experiments were performed in an untreated fused-silica capillary (75/50 cm; 50 μm I.D.) at 10 kV and ambient temperature. The running buffer was 10 mM ammonium acetate solution. It was found that α- and β-CDs did not separate the enantiomers, while γ-CD showed adequate separation capacity. Cu(II) and Zn(II) also improved the chiral separation of NQs, but Fe(II) was ineffective.

a)   b)

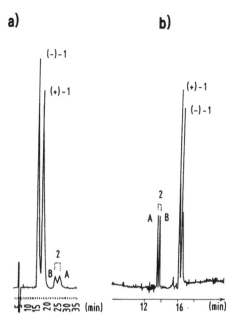

Figure 5.23 *Chiral separations. (a) HPLC, (b) CZE after incubation of racemic flezelastine*
(1) with human liver microsomes. 2A: from (+)-1, 2B: from (−)-1
(Reprinted with permission from ref. 92)

Interestingly, aromatic amino acids considerably enhanced the resolution, but the effect of other amino acids was less. It was assumed that the chiral resolution depends on both the formation of CD NQs inclusion complexes and the ligand exchange interaction. The use of mixed resolution mode was proposed for the enantiomeric separation of other classes of compounds.

The effect of β-CD on the migration of antibacterial sulfonamide derivatives has been studied in detail.[97] Separation of 13 sulfonamides was carried out in a fused-silica capillary thermostated at 25 °C at 20 kV. Drugs were detected with a high-speed scanning multiwavelength UV-Vis detector. Background electrolytes were phosphate–borate buffer solutions adjusted to different pHs and containing various mobile-phase additives. Migration of sulfonamide derivatives decreased with increasing pH of the buffer in the pH range 5.5–7.3, the effect being different for each sulfonamide. This finding indicated that the separation of the drugs can be modified by the appropriate change of pH. The addition of methanol and ACN also influenced the retention and modified the selectivity of separation. Except for sulfathiazole, the mobility of sulfonamides decreased with increasing concentration of β-CD. The formation constants (K_f) and the electrophoretic mobility of the inclusion complexes (μ_{SCD}) were compiled. The forming constants (Table 5.5) show high diversity, drugs with higher pK_a values formed weak complexes with β-CD. It was established that pH, organic modifier and β-CD equally influenced the migration time and migration order of sulfonamide

Table 5.5 *Formation constants (K_f) of inclusion complexes between sulfonamides and β-CD at pH 6.85*

Sulfonamides	K_f	$\mu_{S\text{-}CD}$
Sulfathiazole	1100 ± 4	-0.13
Sulfamethazine	50 ± 3	-0.03
Sulfamethoxypyridazine	560 ± 4	-0.29
Sulfisomidine	60 ± 2	-0.05
Sulfamerazine	270 ± 3	-0.47
Sulfameter	260 ± 2	-0.56
Sulfadiazine	200 ± 2	-0.70
Sulfaquinoxaline	310 ± 3	-0.94
Sulfamonomethoxine	240 ± 2	-0.96
Sulfadimethoxine	140 ± 3	-1.00
Sulfachloropyridazine	560 ± 4	-0.29
Sulfamethoxazole	200 ± 3	-1.30
Sulfisoxazole	190 ± 3	-1.23

K_f in units of M^{-1}; $\mu_{S\text{-}DC}$ in units of 10^{-4} cm^2 V^{-1} s^{-1}.
(Reprinted with permission from ref. 97)

derivatives; therefore, they can be used for the enhancement of the efficiency of separation.

Various CDs have been used for the enantiomeric separation of the antimalarial agent mefloquine.[98] The best separation were achieved with DM-β-CD (Figure 5.24). Resolution increased with increasing concentration in the lower concentration range, but further increases in the concentration did not improve the enantiomeric separation. The reproducibility of migration times was good in normal (0.70–0.78%) and fast separation modes (0.6–1%); however the reproducibility of peak areas was markedly better for normal (0.76–1.69%) than for fast separation mode (20–24%). The sensitivity of the method was 1.2 10^{-6} M for each enantiomer. Calibration was linear in the concentration ranges 2–180 μM (*threo*-mefloquine) and 25–250 μM (*erythro*-mefloquine), the coefficients of correlation being 0.9992. The optimised method was proposed for the determination of mefloquine enantiomers in pharmaceutical preparations.

The enantiomers of Z11556A (Figure 5.25), an intermediate in the synthesis of triazole derivatives with antimycotic activity, have been separated by CE using CDs as chiral selectors.[99] It was found that the best separations could be obtained with native β-CD at 10 mM concentration. The best separation was obtained by using 2-propanol and tetrahydrofuran (THF) organic modifiers; the effect of methanol, ethanol, *n*-propanol, *n*-butanol, *i*-butanol, *tert*-butanol and ACN was markedly lower. It was found that the concentrations of buffer and organic modifier considerably influenced the resolution. Decreasing the applied voltage increased migration time and improved resolution. Native CDs were applied for the enantiomeric separation of muscarinic antagonists (Figure 5.26).[100] Resolution

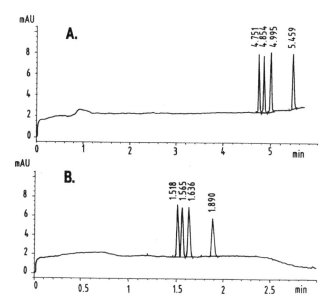

Figure 5.24 *(A) Electropherogram of the enantiomeric separation of the four optical isomers of mefloquine. The peaks at 4.751 and 4.854 represent the threo-mefloquine while at 4.995 and 5.459 the erythro-isomer. Experimental conditions: capillary, 48.5 (39.5 effective length) cm × 50 µm ID; background electrolyte 100 mM phosphate buffer pH 2.5, β-CD concentration 2.5 mM; applied voltage 30 kV; injection by pressure 20 mbar 10 s of racemic threo- and erythro-mefloquine (0.01 and 0.02 mg mL⁻¹, respectively); temperature at 25 °C. (B) Electropherogram of the fast enantiomer separation of threo- and erythro-mefloquine. The background electrolyte, 0.1 M phosphate buffer and 2.5 mM DM-β-CD rinsed the capillary at 50 mbar for 90 s; injection at 20 mbar for 5 s. Threo-mefloquine (1.518, 1.565 min); erythro-mefloquine (1.636, 1.890 min). The distance injection end of the capillary detector was 9 cm. For other conditions as (A)*
(Reprinted with permission from ref. 98)

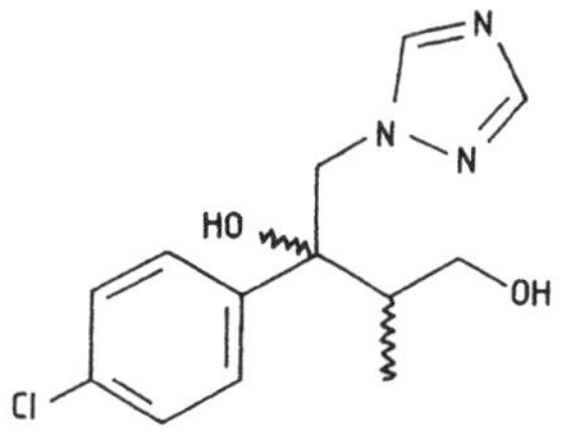

Figure 5.25 *Structure of Z11556A*
(Reprinted with permission from ref. 99)

increased with increasing concentration of CD in the electrolyte but the resolution decreased or remained constant depending on the character of the guest molecule. It was emphasised that separations can only be performed in coated capillaries because the cationic quaternary ammonium groups interact with the silica surface by electrostatic forces.

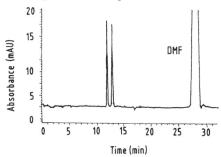

Figure 5.26 *CZE separation of muscarinic antagonists in 25 mM Tris-acetate buffer (pH 4.2). Conditions; capillary, coated, 100 μm ID, 67 cm (59 cm to the detector); injection, electrokinetic mode (1 s, 20 kV); sample, 0.7 mM in water; applied potential, 20 kV; detector UV, 214 nm. The separation was carried out at 25 °C. Trace 1; 2-phenyl-5-[(dimethylamino)methyl]-isoxazolidin-3-one-methiodide salt (25 mM β-CD); Trace 2; 3-(1-hydroxybenzyl)-5-[(dimethylamino)methyl]-isoxazole methiodide salt (10 mM β-CD); Trace 3; 2-benzyl-5-[(dimethylamino)methyl]-isoxazolidin-3-one-methiodide salt (30 mM γ-CD); Trace 4; 3-(1-hydroxy-1-cyclohexylbenzyl)-5-[(dimethylamino)methyl]-isoxazole methiodide salt (20 mM γ-CD); Trace 5; 3-(1-hydroxy-1-phenylbenzyl)-5-[(di- methylamino)methyl]- γ²-isoxazoline methiodide salt (15 mM γ-CD)* (Reprinted with permission from ref. 100)

The chiral separation of catecholamines (norepinephrine, epinephrine, DOPA, and their precursors phenylalanine and tyrosine) was performed by using negatively charged sulfated β-CD (SCD) as chiral selector.[101] The concentration of SCD exerted a decisive influence on the enantiomeric separation (Figure 5.27). The electrophoregrams indicated that a higher concentration of SCD shortened migration time and improved resolution. Because of the good enantioselectivity SCD has been proposed as a chiral selector in CE.

A series of phenethylamines have also been separated by acetylated β-CDs.[102] Enantiomeric separations were carried out in a coated capillary and room temperature. Running buffer consisted of 100 mM potassium dihydrogen phosphate (pH 7.5) containing β-CD, heptakis-(6-O-acetyl)-β-CD (6Aβ-CD) and heptakis-(2,3-di-O-acetyl)-β-CD (Diac-β-CD) at various concentrations (Table 5.6). NMR measurements indicated that the phenyl group of analytes entered the CD cavity and the side chains interacted with the secondary hydroxyl groups on the outer rim of the cavity. The best chiral separations were achieved by Diac-β-CD, but it formed the weakest complexes with the analytes.

Oxedrine, warfarin and other bioactive compounds have been used as test solutes for the selection of optimal conditions for the chiral separation using CDs.[103] TM-β-CD has been employed for the separation of clomiphene Z–E isomers showing estrogenic and antiestrogenic effects.[104] The method was optimised by measuring the resolution at three pH values, buffer, methanol and TM-β-CD concentrations, and injection times (Figure 5.28). The electropherograms emphasise the decisive role of TM-β-CD in the separation, and furthermore prove the applicability of the optimisation process.

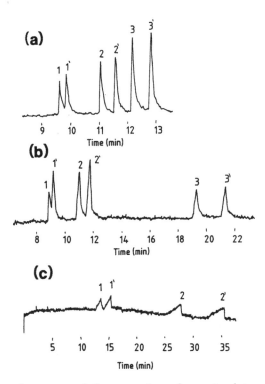

Figure 5.27 *Electropherograms of the separation of norepinephrine (1,1'), synephrine (2,2'), and 2,2-dimethyl-1-phenyl-1-propanol (3,3') in 10 mM phosphate buffer (pH 3.2) with 9.0 mM (a) with 2.3 mM (b) and 0.28 mM of sulfated CDs (c)* (Reprinted with permission from ref. 101)

Table 5.6 *Resolution of various phenethylamines using different cyclodextrins under optimised CE conditions*

Compound	β-CD	6Ac-β-CD	2,3-Diac-β-CD
Norephedrine	–	–	1.07
Ephedrine	0.35	–	0.78
Methylephedrine	0.35	–	1.16
4-Hydroxynorephedrine	<0.3	–	1.30
4-Hydroxyamphetamine	<0.3	<0.3	1.95
Oxilofrine	0.46	–	1.18
Pholedrine	<0.3	<0.3	0.55
Oxedrine	–	–	0.95
Norfenefrine	–	–	0.40
Etilefrine	<0.3	–	2.21
Orciprenaline	0.33	0.31	0.93
Terbutaline	0.54	0.36	0.87
Noradrenaline	–	–	–
Isoprenaline	–	–	–
Salbutamol	–	–	<0.3

(Reprinted with permission from ref. 102)

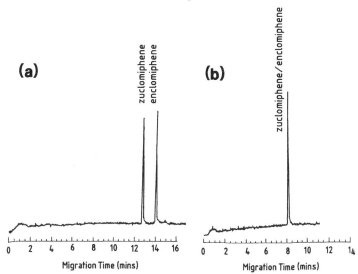

Figure 5.28 *Optimised (a) electropherograms of zuclomiphene (Z isomer of clomiphene) and enclomiphene (E isomer of clomiphene) obtained under the following conditions; 100 mM phosphate buffer, pH 2.3 and 5 mM TM-β-CD. Other conditions; 72 cm × 50 μm ID. capillary (50 cm to detector), UV detection at 254 nm, +30 kV applied voltage, 1.7 s hydrodynamic injection time and temperature 30 °C. (b) Electropherogram showing zero resolution of zuclomiphene and enclomiphene when no TM-β-CD is added. Other conditions are as given in (a)*
(Reprinted with permission from ref. 104)

The enantioseparation of various racemic drugs (sympathomimetic, sympatholytic *etc.*) was successfully achieved by using succinyl β-CD as chiral selector (Table 5.7).[105] It was assumed that the formation of hydrogen bonds between the hydroxyl group of the drugs and the polar substructures on the outer sphere of the CD cavity played a decisive role in the chiral discrimination. Furthermore, cationic analytes can bind to the negatively charged CD by electrostatic forces.

Propranolol and other weak acid and weak basic enantiomers have been also separated by using a single isomer of heptakis-(2,3-dimethyl-6-sulfato)-β-CD (HDMS-β-CD) as chiral selector (Figure 5.29).[106] Nitromethane was employed as external mobility marker. Electroosmotic flow and the effective mobility of analytes depended on the concentrations of both HDMS-β-CD and methanol in the running buffer. It was established that good enantiomeric separation could be obtained for neutral, weak acidic, weak basic and zwitterionic analytes. However, short separation times were obtained only for weak bases migrating cationically.

The enantiomeric separation of seven basic drugs has been performed by CE using CM-β-CD, DM-β-CD and HP-β-CD as chiral selectors.[107] It was established that the best separations could be obtained at acidic pH using CM-β-CD as chiral selector (Figure 5.30).

Table 5.7 *Chiral resolution of the drugs investigated*

Compound	t_l *(min)*	α	R_s
RS-Methoxamine	31.05	1.0431	1.072
RS-Homatropine	38.73	1.0524	1.367
RS-Chlortalidone	40.67	1.5041	7.564
RS-Alprenolol	41.02	1.0334	0.890
RS-Octopamine	41.78	1.0835	1.433
RS-Norephedrine	45.87	1.0889	1.422
RS-Ephedrine	55.24	1.1462	1.189
RS-Propranolol	56.44	1.0317	0.992
RS-Synephrine	58.18	1.0339	0.559
RS-Ethylephrine	72.79	1.0764	0.953
RS-Trihexyphenidyl	78.93	1.0228	0.882

(Reprinted with permission from ref. 105)

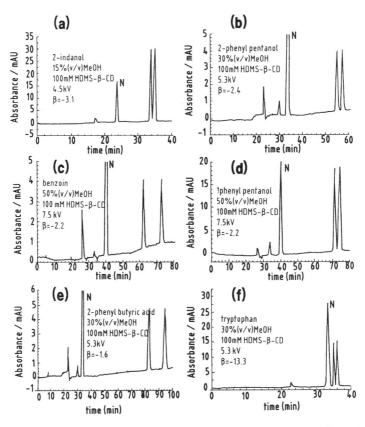

Figure 5.29 *Typical electropherograms of analytes which migrate anionically in the acidic HDMS-β-CD BGEs. The numbers next to the electropherograms indicate the C_{MeOH} (%, v/v), the $C_{HDMS-\beta-CD}$ (mM), the effective applied potential (kV) and the normalised EOF flow mobility (β) values. N, nitromethane marker* (Reprinted with permission from ref. 106)

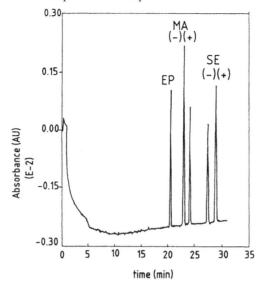

Figure 5.30 *Electropherogram of the chiral separation of a mixture of compounds under the best conditions found for methamphetamine (MA) and selegiline (SE). Ephedrine (EP) was not resolved in this run. Run conditions; Tris-phosphate buffer (pH 2.5; 0.1 M) with EP-β-CD (200 mg mL⁻¹); T = 20 °C* (Reprinted with permission from 108)

A neutral β-CD polymer (EP-β-CD) successfully separated the enantiomers of selegiline, methamphetamine and ephedrine.[108] Because of the good enantioselectivity the method was proposed for the determination of the chiral purity of methamphetamine, selegiline and ephedrine.

The chiral metabolite of the antipsychotic drug haloperidol was separated by using CDs as chiral selectors.[109] It was found that the type and concentration of CD in the electrolyte markedly influenced the enantioselectivity of the system and better separation was achieved at 2.5 than at 4.5 pH. DM-β-CD has also been applied as chiral selector for the enantiomeric separation of phenylalkylamine derivatives (Figure 5.31).[110] The running buffer consisted of 20 mM tris-phosphate (pH 2.7) containing 0.5% hydroxypropylmethyl cellulose and various concentrations of DM-β-CD. The electrophoretic mobility of drugs decreased in the presence of DM-β-CD, the effect depending on the chemical structure of the individual drugs. Binding constants were considerably lower in the presence of 10 and 20% methanol and the optimal concentration of DM-β-CD increased accordingly. It was stated that this simple and rapid CE method can be employed not only for the chiral separation of phenylalkylamine enantiomers but also for the determination of binding constants, facilitating the calculation of the optimal concentration of chiral selector.

A wide variety of other CE methods have been developed and employed for the chiral separation of amphetamine and related compounds. Thus, the chiral

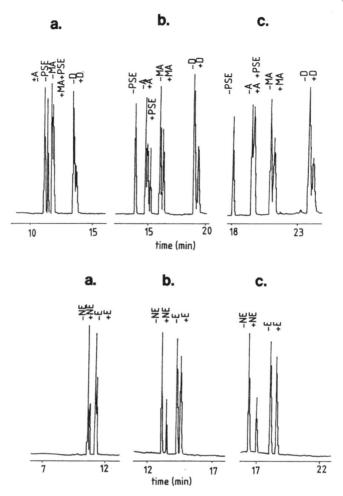

Figure 5.31 *Separation of enantiomers of phenylalkylamine drugs. Upper lane: amphetamine (A), methamphetamine (MA), pseudoephedrine (PSE) and deprenyl (D). Lower lane; norephedrine (NE) and ephedrine (E). Running buffer containing 24 mM DM-β-CD*
(Reprinted with permission from ref. 110)

resolution of 42 enantiomeric pairs (among them amphetamine) has been reported by using hydroxypropyl- and sulfobutyl ether-β-CD.[111] Similar mixtures of neutral and anionic CD derivatives have found application in the enantiomeric separation of cationic drugs such as amphetamine, methamphetamine, cathinone, metcathinone, cathine, cocaine, propoxyphene, and α-hydroxyphenethylamines.[112] Amphetamine and related compounds have been measured in urine after preconcentration by SPE. It was established that the best chiral separations were obtained by hydroxypropyl-β-CD or β-CD and urea mixtures.[113] Another CE technique employed DM-β-CD for the enantiomeric separation of amphetamine.[114] The comparison of five β-CD derivatives used for the enantiomeric resolution of amphetamine and related drugs proved that native β-CD and carboxymethyl-β-CD were the most efficient chiral selectors for this class of analyte.[115] Seven methylated β-CD derivatives have been tested for their capacity to resolve amphetamine and other basic drugs. The results indicated that both the number and position of the methyl substitution considerably influence enantioselectivity.[116] Cyclodextrin-modified CE has been applied for the separation of other basic drugs.[117]

Thus, the chiral resolution of metaproterenol and isoproterenol (chiral selector; DM-β-CD),[118] β-hydroxyphenylethylamine, nor-phenylephrine, terbutaline, ephedrine, norephedrine, ketamine, epinepherine and propanolol (chiral selector; modified β-CD polymer),[119] chlorpheniramine and terbutaline (chiral selectors: β-, hydroxypropyl-β-, and methyl-β-CD),[120] bupivacaine (chiral selectors: modified CDs),[121] carvedilol, tetryzoline, tropicamide and zopiclone (chiral selector; β-CD),[122] ephedrine, propranolol, atenolol, betaxolol and dipivefrin (chiral selectors: native and modified β-CDs,[123] ephedrine and pseudoephedrine (chiral selector; sulfobutyl ether β-CD),[124,125] dimethindene (chiral selectors: native and substituted CDs),[126] denopamine (chiral selector; DM-β-CD),[127] fenfluramine and *meta*-fenfluramine (chiral selectors; β-, γ- and TM-β-CD)[128] and tocaidine and related compounds (chiral selectors: native, hydroxypropyl, hydroxyethyl and trimethyl CDs).[129]

DM-β-CD has also been successfully applied for the enantioseparation of the basic drug ropivacaine by CE-MS.[130] Background electrolyte was 1 mM formic acid pH 2.85 containing 50 mM DM-β-CD. This system allowed the separation of chiral additive and analyte before the MS detector, therefore it can be employed in any CE analysis where the buffer additive exerts a detrimental effect on MS detection.

Enantiomers of the antihypertensive drug pinacidil have also been separated using CDs as chiral additives (Figure 5.32).[131] The results indicated that higher buffer concentrations and the addition of HPC markedly improved the separation. The simple, easy to carry out CE method was proposed for the separation of pinacidil enantiomers.

CDs have been applied for the enantiomeric separation of α_1-adrenoreceptor antagonists and the results were compared with those obtained by HPLC (Figure 5.33).[132] CE measurements were performed in a fused-silica capillary, with a background electrolyte of 0.1 M phosphoric acid pH adjusted to 3.00 with triethanolamine. HPLC separations were carried out on chiral stationary phases (immobilised human serum albumin and α_1-acid glycoprotein). The mobile phase consisted of 1-propanol–0.1 M potassium phosphate buffer, pH 7.4 (10:90, v/v). The analytes were well separated from each other in the presence of β-CD in the running buffer; however, the baseline separation of each enantiomer was not achieved. The results of separation indicated that the best enantiomeric separations could be obtained by using HP-β-CD as a chiral selector. The correlation coefficients were 0.9998 and 0.9996 for the $(-)$- and $(+)$-enantiomers, respectively. Relative standard deviations of peak areas were between 0.6 and 1.8% and

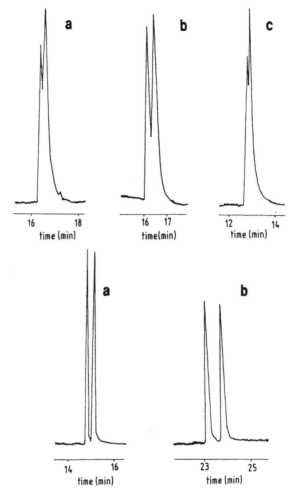

Figure 5.32 *Electropherograms of the enantioseparation of pinacidil. Upper part: background electrolyte, 50 mM Tris-H₃PO₄ (pH 2.3) containing 12 mM β-CD (a), 12 mM HP-β-CD (b) and 12 mM γ-CD (c). Lower part; 100 mM Tris-H₃PO₄ (pH 2.3) containing 9 mM HP-β-CD (a), 9 mM HP-β-CD and 0.05% HPC (b). Conditions; separation tube, 62 cm (41 cm to detector) × 50 μm ID; running voltage, 22 kV; detection 210 nm 0.005 AUFS); temperature, ambient (about 15 °C)*
(Reprinted with permission from ref. 131)

the detection and quantitation limits were 0.2 and 0.6 $\mu g\,mL^{-1}$. It was stated that HPLC analysis of α_1-adrenoreceptor antagonists on protein-coated columns can be employed for the study of the binding properties of the analytes to the corresponding proteins.

The enantioseparation of the 5-hydroxytryptamine receptor antagonist ondansetron was also carried out by using CDs as chiral selectors.[133] Interestingly, α-,

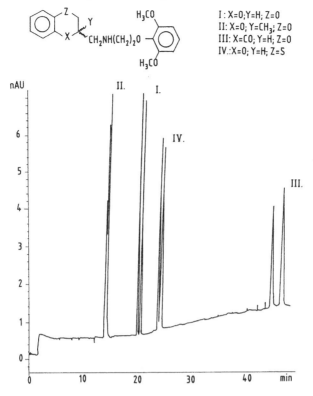

I : X=O;Y=H; Z=O
II: X=O; Y=CH₃; Z=O
III: X=CO; Y=H; Z=O
IV.:X=O; Y=H; Z=S

Figure 5.33 *Separation of enantiomers of compounds (I–IV) LY 213829 derivatives by CZE with β-CD. Conditions; 7.5 mM β-CD in 100 mM phosphoric acid adjusted to pH 3.0 with triethanolamine; separation tube, 48.5 cm (effective length 40 cm) × 50 μm ID; applied voltage, +25 kV; detection 220 nm; temperature, 15 °C; injection time of the sample solution, 10 s*
(Reprinted with permission from ref. 132)

β- and γ-CD, TM-β-CD, HP-β-CD, HP-α-CD, HP-γ-CD and HE-β-CD were not suitable for the separation of ondansetron enantiomers. DM-β-CD showed better chiral separation capacity than CM-β-CD. No interference of the drugs doxorubicin, idarubicin, cyclophosphamide, dacarbazine, cytarabine and etoposide was observed, showing the good specificity of the method. The coefficients of determination were in each instance over 0.999, indicating the good linearity of the calibration between 15 and 250 ng mL⁻¹. Because of the high selectivity and sensitivity the method was proposed for the enantioselective analysis of ondansetron in serum.

The geometric isomers of a leukotriene antagonist have also been separated by CE using CDs as buffer additives and the results were compared with those obtained by HPLC.[134] The analysis time was very long on the coupled C_{18} columns (330 min) and it was markedly reduced on porous graphitised carbon column (PGC) (35 min). Furthermore, isomers were better separated on PGC than on C_{18} columns. The addition of methyl-β-CD, HP-β-CD or α-CD (Figure 5.34) equally resulted in the separation of isomers. The results clearly showed that CE considerably enhanced the resolution of isomers compared with HPLC methods, and therefore its application is highly recommended.

The enantioselectivity of HP-α-CD in CE was investigated in detail using a large set of racemic drugs as model compounds (Table 5.8).[135] Separations were performed in polyacrylamide-coated fused-silica capillaries thermostated at 25 °C. The running buffer consisted of 100 mM sodium dihydrogenphosphate; pH adjusted to 2.5. HP-α-CD was added to the buffer at 30 and 45 mM concentrations. It was stated that the migration retardation factor is related to the stability of the inclusion complexes and it can be used for the optimisation of the enantiomeric separation of any analytes.

Enantiomers of homatropine have also been separated by using β-CD as a chiral additive.[136] Baseline separation of the enantiomers of oxomemazine and tetryzoline has also been obtained by using α-CD as chiral discriminator.[137] The enantioselective separation of tioconazole using hydroxypropyl-β-CD has been previously reported[138] and the thermodynamic parameters of tioconazole CD binding have been determined.[139]

The enantiomeric separation capacity of anionic CD derivatives has been compared for promethazine and other neutral and basic racemic drugs.[140] Optimal enantiomeric separations for trimipramine, trimebutine and chlorprenaline were

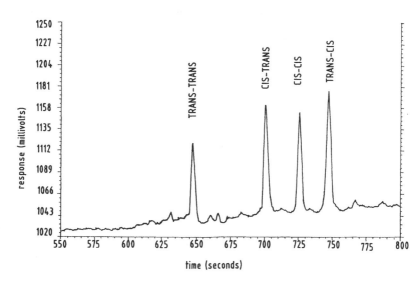

Figure 5.34 *Isomer mixture S(+) and R(−) ondansetron separation by CE. Capillary: 70 cm × 0.05 mm ID fused-silica (62 cm to detector), 30 °C. Run buffer: 0.5 mg mL⁻¹ α-cyclodextrin in 0.1 M disodium tetraborate, pH 9. Voltage: 30 kV. Detector: 238 nm*
(Reprinted with permission from ref. 134)

Table 5.8 *Migration times and parameters derived thereof, for 86 analytes*

Compound name	$t_{m(plain)}$	$t_{m(1)}$	$t_{m(2)}$	α_m	R_m
Alimemazine	5.18	11.92	12.13	1.018	2.342
Aprenolol	5.90	16.64	17.23	1.035	2.920
Amorolfine	5.54	14.37		1	2.580
Atenolol	6.14	7.67		1	1.249
Atropine	5.27	8.60	8.73	1.015	1.657
Azelastine	6.68	15.53		1	2.325
Baclofen	5.20	16.82	17.31	1.029	3.329
Bamethan	5.42	9.76		1	1.801
Benproperine	5.50	10.81	10.94	1.012	1.989
Benserazide	5.20	5.62		1	1.081
Biperiden	7.01	14.96		1	2.134
Bisoprolol	6.81	10.97		1	1.611
Bromphenamine	2.71	8.48	8.57	1.011	3.162
Bupivacaine	4.81	7.60	7.68	1.011	1.597
Bupranolol	5.49	12.57		1	2.290
Butamirate	5.41	17.29		1	3.196
Butetamate	5.04	17.08		1	3.389
Carazolol	5.66	10.88	10.99	1.010	1.942
Carbuterol	5.81	7.41		1	1.275
Carteolol	5.10	7.82		1	1.533
Carvedilol	7.12	15.58		1	2.188
Celiprolol	7.05	11.52		1	1.634
Chloroquinone	2.81	5.24		1	1.865
Chlorphenamine	2.56	7.69		1	3.004
Chlorphenoxamine	4.47	14.93		1	3.340
Cicletanine	5.50	14.35	14.67	1.022	2.667
Clenbuterol	6.35	7.28	7.45	1.023	1.173
Clinidinium bromide	6.42	10.59		1	1.650
Clobutinol	4.85	17.75	17.98	1.013	3.707
Dimetindene	3.17	4.94		1	1.558
Dipivefrine	6.10	10.37		1	1.700
Disopyramide	3.96	6.37		1	1.609
Dobutamine	5.34	9.42		1	1.764
Doxylamine	2.53	4.58	4.74	1.035	1.874
Fendiline	6.05	11.77		1	1.945
Flecainide	5.27	12.19		1	2.313
Gallopamil	6.27	10.64		1	1.697
Homatropine	5.05	7.67	7.91	1.031	1.566
Ipratropium bromide	5.41	8.96		1	1.656
Isoprenaline	5.1	7.18		1	1.405
Isothipendyl	4.80	11.04		1	2.300
Ketamine	3.76	7.13	7.36	1.032	1.957
Meclozine	6.62	17.82		1	2.692
Mefloquine	6.31	11.67	13.39	1.149	2.122
Mepindolol	4.68	8.14	8.26	1.015	1.765
Mequitazine	6.16	14.22		1	2.308
Metaclazepam	5.67	9.78	10.32	1.055	1.820
Metipranolol	6.11	11.21	11.48	1.024	1.879
Naftidrofuryl	6.15	10.63	10.91	1.026	1.774

(continued overleaf)

Table 5.8 *(continued)*

Compound name	$t_{m(plain)}$	$t_{m(1)}$	$t_{m(2)}$	α_m	R_m
Nefopam	4.90	8.96	9.29	1.037	1.896
Nicardipine	6.81	11.06	11.37	1.028	1.670
Norfenefrine	3.95	5.41	5.48	1.013	1.387
Ofloxacin	6.42	8.00	8.37	1.046	1.304
Orciprenaline	5.17	6.97		1	1.348
Ornidazole	13.38	23.91		1	1.787
Orphenadrine	4.46	10.43	10.90	1.045	2.444
Oxomemazine	5.28	9.69		1	1.835
Oxprenolol	5.17	10.77	10.95	1.017	2.118
Oxybutynin	5.43	12.51		1	2.304
Phenoxybenzamine	4.63	9.37	9.84	1.050	2.125
Phenylpropanolamine	3.94	6.21	6.40	1.031	1.624
Pholedrine	4.10	6.39		1	1.559
Pindolol	4.55	7.78	7.95	1.022	1.747
Pirbuterol	3.83	4.00		1	1.044
Prilocaine	6.01	8.05	8.32	1.034	1.384
Procyclidine	3.93	16.35	16.67	1.020	4.242
Promethazine	5.06	12.29	12.41	1.010	2.453
Propafenone	5.42	12.78		1	2.358
Propranolol	4.73	10.55		1	2.230
Reproterol	7.21	8.94		1	1.240
Salbutamol	5.42	6.88		1	1.269
Sotalol	5.33	6.58		1	1.235
Sulpiride	4.71	6.32		1	1.342
Synephrine	4.26	6.39		1	1.500
Talinolol	6.99	15.29		1	2.187
Terbutaline	5.48	7.06		1	1.288
Tetryzoline	4.24	6.78	6.85	1.010	1.616
Theodrenaline	6.96	8.45		1	1.214
Tioconazole	5.48	15.18		1	2.770
Tocainide	4.50	5.99		1	1.331
Trihexyphenidyl	6.23	13.57	13.77	1.015	2.210
Trimipramine	5.49	11.82		1	2.153
Tropicamide	4.73	8.43		1	1.782
Verapamil	6.74	12.98		1	1.926
Zopiclone	6.98	10.33	10.64	1.030	1.524

$t_{m(plain)}$: migration time in plain phosphate buffer; $t_{m(1)}$: migration time of the first eluted enantiomer in 45 mM HP-β-CD containing buffer; $t_{m(2)}$: migration time of the second eluted enantiomer in 45 mM HP-β-CD containing buffer; α_m: migration separation factor; R_m: migration retardation factor. (Reprinted with permission from ref. 135)

determined (Figure 5.35). It was found that anionic CD derivatives are valuable buffer additives for the enantiomeric separation of a wide variety of racemic drugs and they can be used for the chiral separation of drugs, even in blood plasma.

The binding of verapamil enantiomers to CDs has been studied by CE, NMR and electrospray MS and the binding constants have been calculated.[141] Affinity

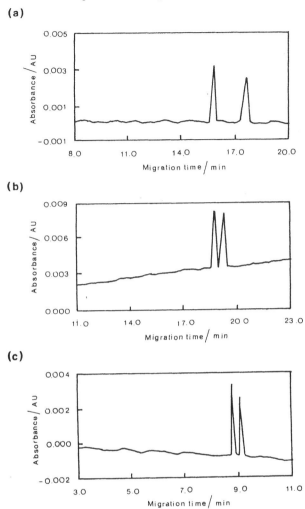

Figure 5.35 *Separation of racemic trimipramine (a), trimebutine (b), and chlorprenaline (c) with charged γ-CD derivatives. Running solution; (a) 5 mM γ-CD phosphate in 50 mM phosphate buffer (pH 5.0); (b) 5 mM CM-γ-CD in the same buffer as (a); (c) 0.5 mM γ-CD phosphate in the same buffer as (a); applied voltage; (a) −12 kV; (b), (c); 12 kV*
(Reprinted with permission from ref. 140)

capillary electrophoresis has been applied for the determination of the binding constant of salbutamol to native and ethylated CDs.[142] Enantiomers of trimetoquinol hydrochloride have also been separated using six different CDs. The best resolution has been achieved by using 5–7% β-CD polymer in the running buffer.[143] Chiral separation of racemate trimipramine, mianserin and thioridazine has been obtained with β-CD as chiral discriminator using formamide and *N*-methyl- and *N,N*-dimethyl-formamide instead of aqueous buffers.[144] Norephedrine and noradrenaline enantiomers have been resolved in running buffer containing HP-β-CD.[145]

Native, alkyl and charged CDs were used for the separation of the stereo-isomers of the α- and β-adrenergic blocking agent labetalol and their efficiencies were compared.[146] The excellent chiral selectivity of charged CDs was tentatively explained by the supposition that, besides the formation of inclusion complexes, ion-pairing interactions also play a marked role in the effect observed. The method showed good validation parameters. Detector response was linear between 0.0142 and 0.142 g L^{-1}, the coefficients of correlation being 0.998–0.999. The limit of detection was 1 mg L^{-1} for each stereoisomer and RSDs for electro-phoretic mobilities and peak areas were less than 1.6% and 1.7%, respectively. The method was proposed for the routine control of the optical purity of labetalol.

The enantiomers of propanolol a β-blocker were also separated by using CDs as chiral selectors together with achiral modifiers (Figure 5.36).[147] It has been concluded from the data that the improvement of enantiomeric separation and the decrease of migration time may be due to the formation of ternary complexes. Ternary complexes held together with hydrogen bonds are probably more rigid than the binary ones resulting in higher complex stability.

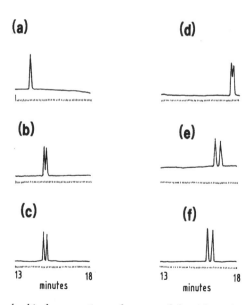

Figure 5.36 *Improved chiral separation of propanolol with and without various co-modifiers. (a) Without β-CD or co-modifier, (b) 40 mM tert-butyl acetate, (c) 40 mM tert-butyl carbamate, (d) 10 mM β-CD and no co-modifier, (e) 40 mM N-(tert-butoxycarbonyl)-glycine, (f) 40 mM tert-butyl(N-hydroxy)-carbamate* (Reprinted with permission from 147)

Native β-CD and seven β-CD derivatives have been employed for the enantiomeric separation of propanolol. It was found that the best chiral separation was achieved by adding carboxymethyl-β-CD to the running buffer.[148] Racemates of terbutaline, terbutaline monosulfate, bambuterol, propanolol, ephedrine and brompheniramine have been resolved by adding various CD derivatives to the running buffer. The results indicated that the number and position of the substituents exerted a considerable effect on the enantioselectivity.[149] The separation of the enantioisomers of propranolol using hydroxypropyl-β-CD as the chiral discriminator has been optimized in a separate study.[150] Enantiomers of propranolol, ephedrin, doxylamine, dimetindene, oxazolidinone and hexobarbital have been separated by using carboxymethyl-, carboxyethyl- and succinylated-β-CDs.[151] Propranolol and another chiral analytes have been further resolved by HP-β-CD,[152] by various modified CDs (HP-, DM- and TM-β-CD[153] and CM- and CE-β-CD[154]).

CE with α-CD chiral selector has been applied for the control of the semi-preparative HPLC purification of the calcium channel blocker amlopidine enantiomers (Figure 5.37).[155] As baseline separation was obtained in the presence of α-CD the method was proposed for the control of the semi-preparative HPLC separation of amlodipine enantiomers.

The enantiomers of carbidopa (a decaroxylase inhibitor) were separated by CE using a single-isomer sulfated β-CD as chiral selector.[156] Background electrolyte consisted of triethanolamine–phosphoric acid (pH 2.5) containing single isomer heptakis-(2,3-diacetyl-6-sulfato)-β-CD (HDAS-β-CD) and polyethyleneglycol 900 (PEG 900). It was stated that the homogeneity of the chiral selector HDAS-β-CD

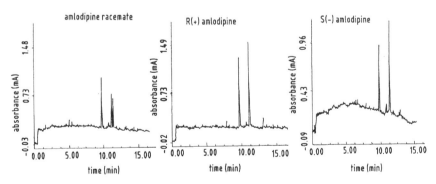

Figure 5.37 *Electropherograms of amlopidine racemate ($0.50\ mg\ mL^{-1}$), R-(+)-amlopidine ($0.52\ mg\ mL^{-1}$) and S-(−)-amlopidine ($0.36\ mg\ mL^{-1}$) preparation (all in form of benzenesulphonate, migration time approx. 9.8 min is benzenesulphonic acid). CE operating conditions: BioFocus 3000 CE instrument, uncoated silica capillary, 50 cm × 50 μm ID, applied voltage 15 kV, detection at a wavelength of 200 nm, injection constant was 68.5 kPa s, electrophoretic buffer; 20 mM Tris-HCl (pH 3.2), 18 mM α-cyclodextrin and 0.05% (m/v) methylcellulose (Reprinted with permission from ref. 155)*

makes possible the development of a reproducible and rapid method for the determination of D-carbidopa impurity in L-carbidopa preparations.

Both CE and MEKC have been employed for the separation of the enantiomers of an acetylcholine esterase inhibitor drug, (R,S)-1-benzyl-4-[(5,6-dimethoxy-1-indanon)-2-yl]methylpiperidine (E2020).[157] It was established that the analyte–CD complex migrated faster than the analyte–micelle complex in MEKC, while the analyte–CD complex migrated more slowly than the uncomplexed analyte in CE. The migration order of enantiomers was different under CE and MEKC conditionsin (Figure 5.38). The limit of quantitation was 3.0 and 0.8% for CE and MEKC, respectively. The principle used for the CE and MEKC separation of enantiomers has been proposed for the analysis of other chiral pharmaceuticals.

Both neutral and charged CDs have been successfully used for the chiral resolution of aminoglutethimide enantiomers the effect of CD mixtures being higher than those of individual CDs.[158] The chiral separation of racemic metomidate and another imidazole derivatives have been achieved by using native CDs, HP and sulfobutyl ether β-CD.[159,160] Enantiomers of *p*-nitrophenyl-2-amino-3-hydroxypropanone, a metabolite of chloramphenicol, have been separated by using β-CD as chiral selector.[161] The use of γ-CD for the enantiomeric separation

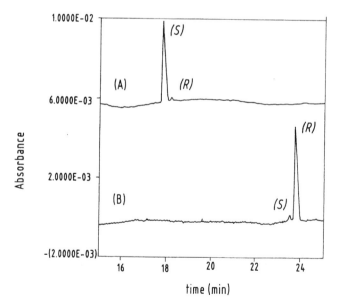

Figure 5.38 *Baseline separation in (S)-E2020 with 3% (R)-E2020 in (A) CZE and (B) MEKC. Conditions: (A) capillary, coated, 47.0 cm (40 cm to the detector) × 50 μm ID; 50 mM phosphate buffer (pH 3.0) containing 5 mM DM-β-CD; applied voltage, 15 kV; (B) uncoated, 46.8 (40.0 cm to the detector) × 50 μm ID; 50 mM phosphate buffer (pH 3.0) containing 65 mM DM-β-CD and 40 mM SDS; applied voltage, 13 kV*
(Reprinted with permission from ref. 157)

of 57 chiral drugs[162] and that of sulfated CDs for 56 compounds has also been reported.[163] The enantiomers of serotonin antagonist tetrahydrocarbazole derivatives have been separated using taurodeoxycholate and β-CD in the mobile phase.[164]

HP-β-CD has been applied for the chiral separation of the enantiomers of the drug (+)-*N*-methyl-γ-(1-naphtalenyloxy)-2-thiophene-propamine hydrochloride.[165]

Separation of Positional and Optical Isomers of Miscellaneous Organic Compounds

Besides environmental pollutants, amino acids and pharmaceuticals, a considerable number of natural bioactive compounds and synthetic products have been separated by chiral CE.

Thus the separation of cytokinin plant hormones in the presence of native CDs and DM-β-CD has been reported.[166] The data indicate that each CD and CD derivative influences the mobility of cytokinins, that is CDs interact with cytokinins. The separation of 15 plant auxin hormones has been achieved by CE using CDs to enhance the efficiency of the method.[167] The separation of derivatised phytosterols has also been facilitated by the addition of DM-β-CD to the running buffer.[168] It was stated that the application of CDs in non-aqueous buffers makes possible the separation of highly hydrophobic and structurally similar compounds such as phytosterols.

CDs together with $MgCl_2$ additive have also facilitated the separation of mono- and dinucleotides of adenosine.[169] The data clearly show that the separation capacity of CE and MEKC is not sufficient without CD additives and the effect of quaternary ammonium salts is negligible. Salts added to the buffer considerably improved separation. CDs influenced the retention; the effect of β-CD was higher than those of γ- and α-CD. The method successfully separated adenine, adenosine and nine adenosine nucleotides in 15 min and was proposed for the analysis of mixtures of adenosine nucleotides. Isomers of adenosine monophosphate have been resolved with β-CD in the running buffer.[170]

CD-modified CE has found application in the analysis of foods and food products. Thus, the successful separation of C_2–C_{18} linear saturated free fatty acids has been obtained by adding TM-β-CD to the running buffer and using indirect absorbance detection.[171] It was established that monosaccharide enantiomers can be separated by CE adding β-CD to the mobile phase.[172] Food preservatives have also been separated by CE using CDs as buffer additives.[173] Food samples were extracted with ethanol or methanol, and the extract was centrifuged and injected in the CE capillary without further prepurification steps. Nine preservatives were separated under optimal conditions (Figure 5.39). Because of the high selectivity, precision and rapidity the method was proposed for the determination of preservatives in foods and food products. Synthetic food colourants[174] and food additive dyestuffs[175] have been determined by CE using β-CD to improve resolution. It has been reported that α-CD and sulfated β-CD

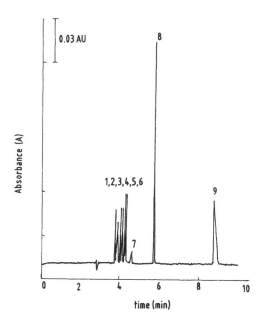

Figure 5.39 *Separation of nine preservatives by CE. Conditions; separation solution, 0.035 M borax–NaOH buffer (pH 10.0) containing 2 mM α-cyclodextrin; capillary, 47 cm (40 cm to the detector) × 50 μm ID; applied voltage 20 kV; detection wavelength, 195 nm; concentration of each analyte, 125 μg mL^{-1}; pressure injection, 3 s; column temperature, 25 °C. Peak identification: 1 = n-butyl p-hydroxybenzoate; 2 = methyl p-hydroxybenzoate; 3 = n-propyl p-hydroxybenzoate; 4 = ethyl p-hydroxybenzoate; 5 = isobutyl p-hydroxybenzoate; 6 = isopropyl p-hydroxybenzoate; 7 = sorbic acid; 8 = benzoic acid, and 9 = p-hydroxybenzoic acid*
(Reprinted with permission from ref. 173)

derivatives improved the separation of neutral, cyclic and bicyclic mono-terpenes.[176]

Positional isomers of nitroaromatic explosives have been separated by the addition of negatively charged CDs to the running buffer in electrokinetic chromatography (EKC) and the results were compared with those obtained by HPLC.[177] The elution order of explosives was similar in HPLC and EKC with SDS indicating that the selectivity of SDS is similar to that of C$_{18}$ HPLC stationary phase. Best separations were achieved by using neutral and negatively charged CDs (Figure 5.40). It was found that the combined application of sulfobutyl ether-β-CD, SDS and ACN facilitated the separation of 3-nitrobenzene/4-nitrobenzene and 1,3-dinitrobenzene/1,4-dinitrobenzene pairs.

The capacity of various β-CD derivatives for the enantiomeric separation of vicinal diols has been studied in detail (Figure 5.41).[178] The background electrolyte consisted of 50 mM borate buffer containing CDs (β-CD, methyl-β-CD, DM-β-CD, TM-β-CD, HE-β-CD, HP-β-CD and succinyl-β-CD). It was found that β-CD, methyl-β-CD, HE-β-CD, HP-β-CD and succinyl-β-CD separated the

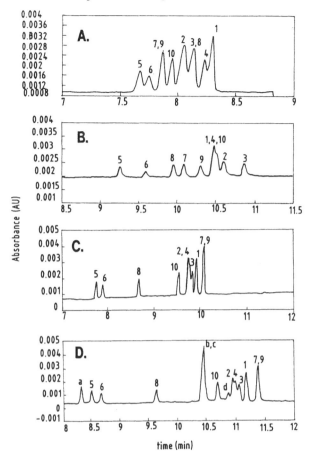

Figure 5.40 *Separation of trinitrotoluene (peak a), 1,3-dinitronaphtalene (DNN, peak b), nitrobenzene (peak c) and 1,5-DNN (peak d) from the above ten explosive model using 20 mM borate buffer pH 9 containing 30 mM sulfobutylether-β-cyclodextrin and 10 mM HP-β-CD, resultant current; 39.8 μA and $t_{eof} = 5.81$ min. Peak identification: 1 = 2-nitrotoluene; 2 = 3-nitrotoluene; 3 = 4-nitrotoluene; 4 = 1,2-dinitrobenzene; 5 = 1,3-dinitrobenzene; 6 = 1,4-dinitrobenzene; 7 = 2,3-dinitrotoluene; 8 = 2,4-dinitrotoluene; 9 = 2,6-dinitrotoluene and 10 = 3,4-dinitrotoluene*
(Reprinted with permission from ref. 177)

vicinal diols only in borate buffer. Methanol added to the running buffer improved chiral separation and increased migration time. It was concluded from the data that borate complexation is a prerequisite of successful enantiomeric separation. The simultaneous use of CDs and borate buffer is an effective tool for the chiral separation of this class of analytes.

Similar method has been employed for the enantiomeric separation of quinazo-line-(3*H*,1*H*)-4-one-2-thiones and tetrazol-5-thiones.[179] The best separations were

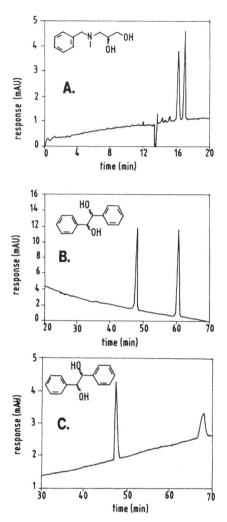

Figure 5.41 *Electropherogram of the chiral resolution of (A) R,S-3-(N-benzyl-N-methyla-mino)-1,2-propanediol. Conditions; 1.8% β-CD, 50 mM borate, pH = 9.3; (B) R,R,S,S-dihydrobenzoin. Conditions; 1.8% β-CD, 50 mM borate, pH = 9.3 and 20% (v/v) methanol. (C) R,R,S,S-dihydrobenzoin. Conditions; 2.0% succinyl-β-CD, 50 mM borate, pH = 9.3*
(Reprinted with permission from ref. 178)

achieved with β-CD (Figure 5.42); α-CD, γ-CD and DM-β-CD were ineffective. However, β-CD separated the analytes only in borate buffer, emphasising the importance of borate complexation in the chiral separation process. It was established that efficacy of separation depends on the pH and on the concentration of β-CD and buffer. The good enantioseparation was explained by the supposition that the bulky substructures of diols entered the CD cavity and the diol group interacted with the borate ions in the buffer.

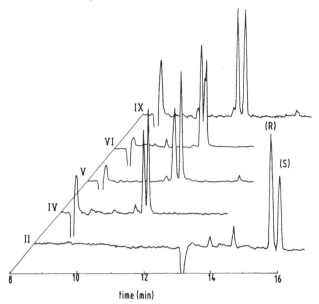

Figure 5.42 *Electrophoregram of the chiral resolution of compounds II, IV–VI and IX. Conditions; II, 1.8% β-CD, 0.1 M Tris and 50 mM borate, pH 9.3, 214 nm, 15 kV; V, IX, 1.8% β-CD, 50 mM borate, pH 9.3, 214 nm, 15 kV; IV, VI, 1.8% β-CD, 100 mM borate, pH 9.3, 280 nm, 20 kV. Peak identification: II = RS-3-Benzyloxy-1,2-propanediol; IV = 3-o-Bromphenyl-2-(2,3-dihydroxy-propylthio-4(3H)-quinazolinone; V = 3-o-Iodphenyl-2-(2,3-dihydroxy-propylthio-4(3H)-quinazolinone; VI = 3-o-Ethylphenyl-2-(2,3-dihydroxy-propylthio-4(3H)-quinazolinone; IX = 1-p-Bromphenyl-5-2,3-dihydroxypropylthiotetrazol* (Reprinted with permission from ref. 179)

Different CDs and CD derivatives have been employed for the enantiomeric separation of chromane compounds in CE.[180] The results clearly showed that γ-CD was the best chiral selector. However, other CDs also have considerable enantiomeric separation capacity in some instances. The optimal concentration of CDs depended highly on the type of CD and the enantiomers to be separated, on the pH and on the concentration of buffer. The RSD of the migration time was 5% and the detection limit for the stereochemical purity of enantiomers was 0.08% (area/area).

CDs together with a non-chiral crown ether (18-crown-6) were employed for the enantiomeric separation of non-polar primary amines.[181] It was found that enantiomers could not be separated with CDs or with 18-crown-6 alone, but their mixture provided baseline separation. It was assumed that the amines formed complexes with the crown ether and the hydrophobic part of the complex entered the CD cavity, resulting in a more rigid complex structure.

Functionalised piperidine enantiomers have also been separated with native, alkylated and anionic β-CDs as chiral selectors.[182] Enantiomers of dienomycine C were partially separated by β-CD and HP-β-CD, addition of methanol to the buffer improved separation. Functionalised piperidine was successfully separated with DM-β-CD. Dienomycine enantiomers were baseline separated by SBE-β-CD as shown in Figure 5.43. It was established that the enantioselectivity of anionic SBE-β-CD was higher than that of neutral CDs, therefore its application in CE for the separation of other pairs of enantiomers has been proposed.

Acids

Positional isomers of hydroxynaphthalenecarboxylic acids (HNs) were separated through the formation of inclusion complexes with β-CD.[183] It was found that 1,2-, 2,1-, 2,3- and 6,2-HNs are baseline separated in buffers pH 10.1 containing 0.5-1 mM β-CD (Figure 5.44).

Native CDs have been employed to improve of separation of aromatic sulfonic acids.[184] The addition of α-CD to the running buffer did not improve the separation, but β- and γ-CDs alone or in combination greatly enhanced the separation capacity of the system. It was further established that not only the concentration of CDs but also the pH exerted a considerable influence on the

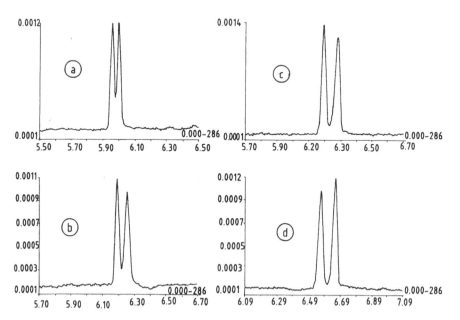

Figure 5.43 *Electropherograms of dienomycin with increasing concentrations of sulfobutyl ether β-CD. Capillary; 44 cm × 50 µm; applied voltage: −15 kV; tempera-ture: 25 °C; detection at 286 nm; electrolyte: 63.5 mM H_3PO_4−46.9 mM NaOH (pH 2.6), hydrodynamic injection time; 1 s; solute concentration; 25 µg mL⁻¹. SBE-β-CD concentration of 20 mg mL⁻¹*
(Reprinted with permission from ref. 182)

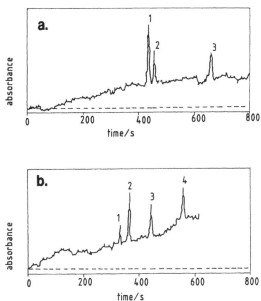

Figure 5.44 *(a) Electropherogram of a mixture of 1,2-HN (HN = hydroxynahphthalene-carboxylic acids) (Signal 1), 2,1-HN (Signal 2), 2,3-HN (Signal 1), and 6,2-HN (Signal 3) in pH 10.1 buffer; (b) Electropherogram of a mixture of 1,2-HN (Signal 1), 2,1-HN (Signal 3), 2,3-HN (Signal 2), and 6,2-HN (Signal 4) in pH 10.1 buffer containing β-CD (5.0 × 10⁻³ mol dm⁻¹)*
(Reprinted with permission from ref. 184)

separation of naphthalenesulfonic acids. Elevated pressure resulted in shorter analysis time without marked loss of separation efficacy (Table 5.9). Separations were better in borate than in phosphate buffers. It was concluded that the method was suitable for the detection of 0.1% or lower impurities in aromatic sulfonic acid preparations. Various aromatic aminosulfonic acids have also been separated with CE using β-CD to improve resolution.[185] Naphthalenesulfonylamino acids have been separated by CE using selectively methylated β- and γ-CD derivatives as chiral discriminators.[186] The influence of ^2H_2O on the formation of inclusion complexes of 2-naphthalenesulfonate (2NS) with β-CD has also been studied by CE and the results were compared with those obtained by UV spectrometry.[187] The migration time of 2NS decreased with increasing concentration of β-CD in the running buffer both in H_2O and in ^2H_2O. It was found that the K value determined in ^2H_2O is 27% higher than that determined in H_2O. It was assumed that the hydrophobic and electrostatic interactions are different in the presence of the ^2H isotope. Spectrometric results entirely supported the conclusions drawn from CE measurements. Naphthalenesulfonic acids have also been separated with randomly sulfated, randomly aminated and heptasulfate β-CDs.[188]

Table 5.9 *Migration times of amino-substituted naphthalene sulfonic acids for the different conditions used*

Overpressure conditions	Buffer	pH	Migration time					
			I	II	III	IV	V	VI
Without overpressure	Borate[a]	7.0	5.8	5.4	5.9	5.3	12.6	6.7
		7.5	5.8	5.8	6.0	5.8	13.7	7.1
		8.0	5.7	5.7	5.7	5.7	11.9	6.7
		8.5	6.3	5.9	6.0	6.3	14.1	7.4
		9.0	9.2	8.8	8.2	9.6	29.5	11.1
	Phosphate[b]	9.5	7.1	6.7	6.6	7.3	19.5	8.5
Overpressure 2000 Pa applied	Borate-phosphate[c]	7.0	5.4	5.3	5.6	5.3	N/A	N/A
		9.0	6.1	6.6	5.9	6.1	N/A	7.2

Overpressure conditions	Buffer	pH	Migration time			
			VII	VIII	IX	X
Without overpressure	Borate[a]	7.0	5.5	5.8	6.6	10.7
		7.5	6.5	6.7	8.0	14.0
		8.0	7.7	7.3	9.2	17.3
		8.5	14.9	10.3	16.9	43.2
		9.0	17.8	10.8	19.1	34.1

Overpressure 2000 Pa applied	Borate[a]	7.0	4.6	4.8	5.4	7.8
		7.5	5.3	5.4	6.2	9.3
		8.0	5.9	5.7	6.8	10.0
		8.5	8.8	7.0	9.4	13.6
		9.0	10.0	7.4	10.4	13.5
Overpressure 2000 Pa applied	Borate-phosphate[c]	8.0	9.03	9.03	11.17	22.91
		8.5	9.09	7.86	10.29	17.32

Voltage, +15 kV; temperature, 35 °C. [a]0.025 mol L^{-1} borate buffer + 0.01 mol L^{-1} β-CD; capillary, 75 cm (60 cm to detector) × 50 μm I.D. [b]0.025 mol L^{-1} phosphate buffer + 0.01 mol L^{-1} β-CD; capillary, 75 cm (60 cm to detector) × 50 μm I.D. [c]0.025 mol L^{-1} borate–phosphate buffer + 0.01 mol L^{-1} β-CD; capillary, 75 cm (60 cm to detector) × 50 μm I.D. I = 7-amino-1-hydroxynaphthalene-3-sulfonic acid; II = 6-amino-1-hydroxynaphthalene-3-sulfonic acid; III = 2-aminonaphthalene-7-sulfonic acid; IV = 1,6-dihydroxynaphthalene-3-sulfonic acid; V = 6-aminonaphthalene-1,3-disulfonic acid; VI = 5,5′-dihydroxy-2,2′-dinaphthylamine-7,7′disulfonic acid; VII = 1-amino-8-hydroxynaphthalene-3,6-disulfonic acid; VIII = 1,8-dihydroxynaphthalene-3,6-disulfonic acid; IX = 8-amino-3-hydroxy-naphthalene-1,6-disulfonic acid; X = 8-aminonaphthalene-1,3,6-trisulfonic acid.
(Reprinted with permission from ref. 184)

Various organic acids have also been resolved by CD-modified CE. Thus, 2-hydroxy acids (chiral selectors: neutral and charged CDs),[189] D-galactonic and D-gluconic acids (chiral selector; β-CD),[190] isomeric benzoic acids (chiral selectors: native CDs)[191] and a set of nine organic acids (chiral selectors; HP-β-CDs with various degrees of substitution)[192] have been successfully resolved.

Not only organic acids but also enantiomers of weak bases have been resolved by CE using CDs as chiral selectors. The simultaneous application of a non-chiral crown ether (18-crown-6) and β-CD for the enantiomeric separation of primary amino compounds has also been reported.[193] The chiral separation of two cyclic amines with β-CD and neutral positional isomers and enantiomers with CM-β-CD has also been obtained.[194]

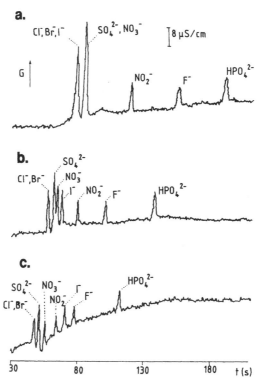

Figure 5.45 *Electropherograms from the separation of inorganic anions at various β-CD concentrations in the carrier electrolyte without α-CD. The concentrations of the anions in the injected samples were 10 μM. The separations were carried out with the driving current stabilised at 30 μA. The voltage between the driving electrodes was 5.5 kV in the carrier electrolyte solution without α-CD. It was 11.0 kV when the carrier electrolyte solution contained α-CD at a 75 mM concentration (for other concentrations it was between 5.5–11 kV in a proportion to the concentration of α-CD). For further details on the composition of the carrier electrolyte solutions see text. G = Increasing conductivity*
(Reprinted with permission from ref. 196)

Anions

CDs have found application in the isotachophoretic separation of inorganic anions in a polytetrafluoroethylene (Teflon) capillary.[195] Anions were detected with a contactless conductivity detector. The effect of the β-CD concentration on the separation of inorganic anions was determined a (Figure 5.45). The electrophoregrams showed that the migration time of anions is shorter at higher concentrations of α-CD in the electrolyte and the anions are better separated. Unfortunately, the separation of Cl⁻ and Br⁻ was not achieved under the experimental conditions. The method has been successfully applied for the analysis of a dietary salt sample and diluted milk. It was established that organic acids did not co-migrate with the anions, influencing their separation and quantitation. Native β-CD, CM-β-CD and sulfoethyl ether and sulfobutyl ether of β-CD have been employed for the chiral separation of R/S-1,1'-binaphthyl-2,2'-diyl hydrogen phosphate by capillary electrophoresis using anionic cyclodextrin derivatives as chiral selectors.[196,197] Atropisomeric binaphthyl derivatives have also been resolved by adding various CD derivatives to the running buffer.[198] A set of 23 racemic compounds (neutral, weak acid, weak base, zwitterionic) has been employed for the study of the enantioselectivity of single isomer chiral resolving agents, such as heptakis-(2,3-diacetyl-6-sulfato)-β-CD,[199] hepta-6-sulfato-β-CD[200] and heptakis-(2,3-dimethyl-6-sulfato)-β-CD.[201]

2 Micellar Electrokinetic Chromatography

Micellar electrokinetic chromatography (MEKC) is a combination of electrophoresis and chromatography. A surfactant (or any other micelle forming compound) is added to the background electrolyte above its critical micellar concentration. The partition of analytes between the hydrophobic micelles and the hydrophobic aqueous buffer is determined by their hydrophobicity. The different partition of analytes between the hydrophobic and hydrophilic phases accounts for the separation. MEKC has been generally used for the separation of neutral hydrophobic molecules. MEKC together with other electrokinetic chromatographic methods without electroosmotic flow has been previously reviewed.[202] Reviews have been devoted to the application of MEKC[203] and electrokinetic chromatography[204] in drug analysis.

Separation of Positional and Optical Isomers of Pesticides and Other Environmental Pollutants

Enantiomers of neutral pesticides have been separated by MEKC.[205] Separation was performed in a fused-silica capillary at constant temperature. Running electrolytes were prepared from 40 mM borate buffer (pH 9.0) and 100 mM sodium dodecyl sulfate (SDS) solutions. Buffers were modified by the addition of methanol, ACN, α-, β-, γ-, HP-β-, DM-β- and TM-β-CD and the effect of additives on the resolution was determined. It was assumed that the analyte migrates with a velocity between those of the electroosmotic flow and the micelles. It was stated that the addition of 20% methanol or 15% ACN considerably increased the efficacy of separation of organophosphorus pesticides, such as malathion, fenamifos, isofenphos, ruelene and dialifos. The same effect was achieved by the addition of 40 mM DM- or TM-β-CD to the running buffer. The simultaneous application of γ-CD, ACN and SDS resulted in the chiral separation of DDT and related compounds (Figure 5.46). Not only the enantiomers of DDT and related compounds but also those of organophosphorus pesticides were separated by MEKC. It was found that the separation capacity of CD-MEKC is sometimes higher than those of the corresponding GC and HPLC methods, but the sensitivity is markedly lower.

Structural homologues of alkylphenols have also been separated by CD-modified MEKC.[206] Separations were carried out in a fused-silica capillary using 12.5 mM borate buffer (pH 9) and 25 mM SDS. It was found that the addition of ACN and HP-β-CD to the running buffer markedly improved the separation of alkylphenols. The concentration and type of CDs, the amount of organic modifier and SDS equally influenced the separation of the isomers of 4-nonylphenol.

The beneficial effect of CDs on the separation of polychlorinated biphenyl (PCB) congeners has also been established.[207] Fused-silica capillaries (50 μm I.D.) with different total and effective lengths have been used for the separations. Electrolytes consisted of 2-(N-cyclohexylamino)ethanesulfonic acid (CHES) buffer (pH 10.0) containing SDS, urea, β- and γ-CD at various concentrations. Capillaries were thermostated at 45 °C. It was established that the retention factor k depended on the inverse of the CD concentration and on the concentration of SDS in the electrolyte, on the critical micelle concentration (CMC) and on the

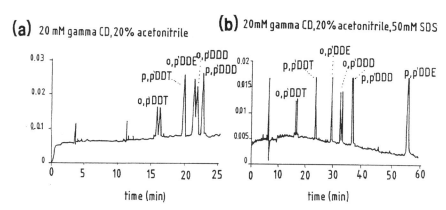

Figure 5.46 *Separation of p,p'-DDT, p,p'-DDD, o,p'-DDE, p,p'-dde and the enantiomers of o,p'-DDT, o,p'-DDD as a function of the addition of γ-CD and acetonitrile to the separation buffer; (a) with 20 mM borate buffer (pH 9.0), 100 mM SDS, 30 °C, 30 kV, (b) with 20 mM borate buffer (pH 9.0), 50 mM SDS, 25 °C, 20 kV*
(Reprinted with permission from ref. 205)

partial specific volume of CD and micelle. The effects of buffer composition on the separation of PCB congeners were demonstrated (Figure 5.47). The best separation was achieved by using a mixture of β- and γ-CD while only poor separation was observed in buffers without CDs. It was assumed that the bulky hydrophobic ring structure of PCBs and the alkyl chain of SDS enter the apolar cavity of CDs forming ternary complexes which facilitate the separation. Although PCB congeners were well separated by CD-MEKC, the detection limit was fairly high ($100-500$ mg L^{-1}) making difficult the analysis of real environmental samples. The mixture of β- and γ-CDs has also been used for the chiral separation of a slightly different set of PCBs.[208] Experiments were carried out in a fused-silica capillary thermostated at 45 °C. The running electrolyte was CHES buffer (pH 10.0) containing SDS and β- and γ-CDs. Separation factors increased with increasing concentration of buffer. Good enantiomeric separation capacity but relatively low sensitivity of the method were reported. PCBs have been further separated by adding γ-CD[209] or 2-hydroxy-γ-CD[210] to the running buffer.

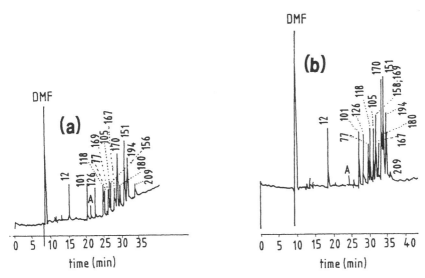

Figure 5.47 *Separation of fourteen PCB congeners by CD-MEKC using 60 mM γ-CD (a) in 0.1 M SDS and using a mixture of 72 mM β-CD and using 25 mM γ-CD in 0.09 M SDS (b). Other buffer conditions; 0.08 M CHES (pH 10.0), 2 M urea. Capillary, 65 cm (50 cm effective length) × 50 μm ID. Injection, 1.2 s, 20 mbar. Voltage, 15 kV. Temperature, 45 °C. Detection, 240 nm. A = unknown peak. Peak identification: 12 = 3,4-dichloro-; 77 = 3,3',4,4'-tetrachloro-; 101 = 2,2',4,5,5'-pentachloro-; 105 = 2,3,3',4,4'-pentachloro-; 118 = 2,3',4,4',5-pentachloro-; 126 = 3,3',4,4',5-pentachloro-; 151 = 2,2',3,5,5', 6-hexachloro-; 156 = 2,3,3',4,4',5-hexachloro-; 167 = 2,3',4,4', 5,5'-hexachloro-; 169 = 3,3',4,4',5,5'-hexachloro-; 170 = 2,2', 3,3',4,4',5-heptachloro-; 2,2',3,4,4',5,5'-heptachloro-; 194 = 2,2',3,3',4,4',5, 5'-octachloro- and 209 = 2,2',3,3',4,4',5, 5',6,6'-decachlorobiphenyl*
(Reprinted with permission from ref. 207)

The use of CD-MEKC for the separation of various polycyclic aromatic hydrocarbons (PAHs) has also been vigorously investigated. Thus, the separation of hydroxylated PAHs (OH-PAHs) has been reported using γ-CD and urea in the running electrolyte (Figure 5.48).[211] The buffers contained 2.5 mM sodium tetraborate, 12.0 mM boric acid (pH 9.0) and various concentrations of γ-CD, SDS and urea. It was concluded that CD-MEKC was suitable for the separation of OH-PAHs, that its efficiency was higher than that of the corresponding HPLC method and it could be employed for the analysis of urine, serum and blood samples. A similar set of hydroxy-PAHs has been separated using a slightly different experimental set-up.[212] The effect of γ-CD concentration on the migration time and separation of OH-PAHs was studied. Detection limits were low $(0.08-0.5) \times 10^{-15}$ M showing the high sensitivity of the detection system. Twenty PAHs were separated using a mixture of β- and γ-CDs as buffer additives.[213] The data showed again that the total amount of CDs, their ratio and the concentration of buffer simultaneously influence the efficiency of separation. It was stated that CD-MEKC offers a valid alternative for the separation of PAHs, although the detection limit is higher than that of other chromatographic techniques, such as GC and HPLC. The fact that a good linear relationship was

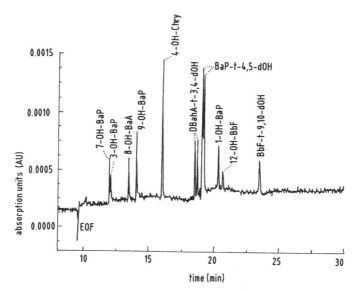

Figure 5.48 *CD-MEKC separation of ten hydroxylated polycyclic aromatic hydrocarbons; analytes: 1-OH-BaP, 3-OH-BaP, 7-OH-Bap, 9-OH-Bap, 8-OH-BaA, 4-OH-Chry, 12-OHBpF, Bap-t-4,5-dOH, BpF-9,10-dOH and DBahA-t-dOH; electrophoretic conditions: 12 mM Na₂B₄O₇, 2.5 mM H₃BO₇, 50 mM SDS, 30 mM γ-CD, 4 M urea; UV detection at 270 nm, 10 kV+). BaP = benzo[a]pyrene; BaA = benzo[a]anthracene; Chry = chrysene; BpF = benzo[b]fluoranthene, and DBahA = dibenz[ah]-anthracene*
(Reprinted with permission from ref. 211)

found between the logarithm of capacity factor and the hydrophobicity of solutes indicates the involvement of hydrophobic interactive forces in the separation mechanism. PAHs have been further resolved with native and derivatised γ-CDs[214,215] and with a mixture of neutral and charged CDs.[216]

Sulfobutyl ether-β-CD (SBE-β-CD) has also found application in MEKC and its effect has been compared with that of SDS for the separation of hydrophobic compounds, including aryl amines, aldehydes, ketone and alkylbenzenes (Figure 5.49).[217] Both SDS and SBE-β-CD were suitable for the separation of these groups of analytes. The differences in the retention order may be due to the formation of inclusion complexes between the analytes and SBE-β-CD. Separation of the isomers of dimethylnaphthalenes has also been obtained by using γ-CD as chiral discriminator.[218]

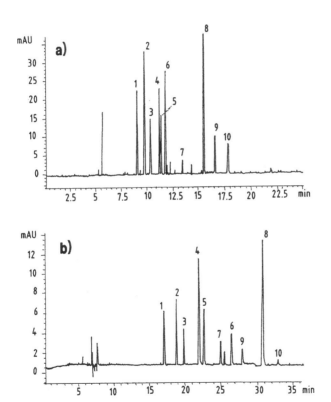

Figure 5.49 *MEKC of aromatic compounds. Buffer; (a) 25 mM borate, pH 9.2, 100 mM SDS, (b) 25 mM borate, pH 9.2, 50 mM SBE-β-CD. Peak identification; 1 = aniline, 2 = benzyl alcohol, 3 = benzaldehyde, 4 = nitrobenzene, 5 = acetophenone, 6 = phenylacetone, 7 = toluene, 8 = chlorobenzene, 9 = ethylbenzene, and 10 = naphthalene*
(Reprinted with permission from ref. 217)

Separation of Positional and Optical Isomers of Miscellaneous Organic Compounds

CD-MEKC has been employed for the analysis of various pharmaceuticals. The enantioseparation of racemic norgestrel by CD-MEKC has been described.[219] It was observed that the migration time of enantiomers increased with increasing concentration of CD and SDS in the running buffer. The method was fast and separated the enantionen well. The chiral separation of racemic diniconazole and uniconazole has been achieved by adding CD derivatives to the running buffer.[220] Chiral resolution of chlorpheniramine has also been obtained using β-CD as chiral discriminator.[221] Phenylethylamine stimulants, the antidepressant diclofensine, β-agonists and β-antagonists have been optically resolved with HP-β-CD and sulfobutyl ether-β-CD.[222] Active components (cinnamaldehyde, cinnamic acid, paeoniflorin, glycyrrhizin and [6]-gingerol) of the medicinal herb preparation Kuei-chih-tang extract have also been separated with CD-MEKC (Figure 5.50).[223] The method showed good validation parameters: the relative standard deviation of migration times varied between 0.25–0.42%, the coefficients of correlation of the calibration were 0.993–0.999, and the detection limit was the highest for oleanolic acid (28.43 μg mL^{-1}) and the lowest for [6]-gingerol (0.77 μg mL^{-1}).

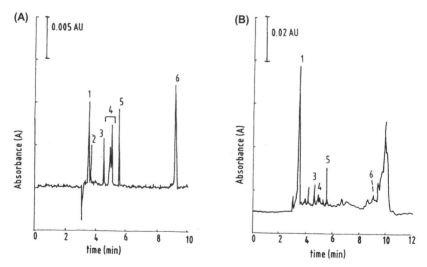

Figure 5.50 *Separation of (A) specific analytes and (B) Kuei-chih-tang extract by CD-MEKC. Conditions: separation solution, 38 mM borax–NaOH buffer containing 20 mM SDS and 2 mM γ-cyclodextrin, pH 10.0; pressure injection, 3 s; capillary, 47 cm (40 cm to detector) × 50 μm ID; applied voltage, 20 kV; column temperature, 25 °C. Detection wavelength for (B); 200 nm (before 4.0 min), 270 nm (after 4.0 min), 295 nm (after 4.8 min), 270 nm (after 5.3 min) and 200 nm (after 8.0 min). Peak identification: 1 = Paeoniflorin; 2 = oleanolic acid; 3 = glycyrrhizin; 4 = cinnamaldehyde; 5 = cinnamic acid; 6 = [6]-gingerol*
(Reprinted with permission from ref. 223)

The method was considered as a promising alternative to other chromatographic techniques applied for the analysis of traditional Chinese medicinal preparations.

The baseline separation of five biflavonoids of Gingko Biloba has been achieved by CD-MEKC.[224] It was established that the DM-β-CD considerably improved the separation of the highly hydrophobic analytes. CD-MEKC has been used for the separation of other bioactive compounds. Dansylated amino acids and other analytes were separated by using vancomycin or CD-modified MEKC.[225] Chiral resolution of di- and tri-peptides has been performed by using β- or γ-CD as chiral selectors.[226] The measurements indicated that SDS also entered the cyclodextrin cavity, competing with the analyte in the formation of inclusion complexes. The competition can modify the migration of analytes. Various CDs and CD derivatives have been employed for the MEKC separation of fat-soluble vitamins and the results were compared with those obtained by HPLC.[227] DM-β-CD made possible the separation of α-, β-, γ- and δ-tocopherols (Figure 5.51). The retention time of vitamins (A, D_2, D_3 and E) was strongly reduced in HPLC by adding DM-β-CD to the mobile phase, the effect of HE-β-CD and triacetyl-β-CD being smaller. The four isomers of vitamin E had not been separated by HPLC even in the presence of CDs. It was concluded from the data that the addition of CD improved the separation of fat-soluble vitamins in both MEKC and HPLC; however, the separation capacity of MEKC was higher than that of HPLC.

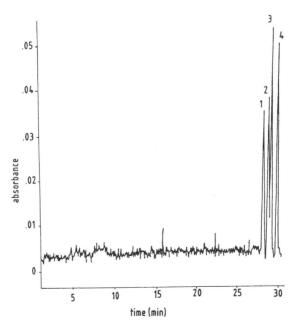

Figure 5.51 *Separation of α-, β-, γ- and δ-tocopherol by CD-MEKC. Conditions 20 mM phosphate, 50 mM borate, 30 mM SDS, 25 mM DM-β-CD, pH 8.0, V = 22 kV* (Reprinted with permission from ref. 227)

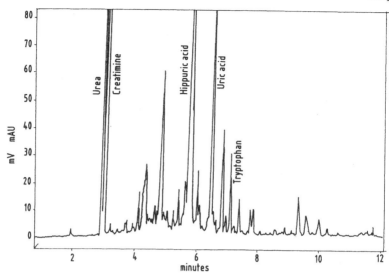

Figure 5.52 *Pooled urine separated using optimised conditions, 30 mM sodium tetraborate, pH 10, 75 mM SDS, 10 mM β-CD, 15 °C, 20 kV, 4 s hydrodynamic injection, detection at 195 nm*
(Reprinted with permission from ref. 228)

The performance of CE and MEKC was compared in the analysis of UV-absorbing components in human urine on a fused-silica column.[228] Urine samples were injected without any special pretreatment. Running electrolytes consisted of sodium tetraborate buffer and phosphate buffer (concentration and pH varying) containing SDS and β-CD in various concentrations. As it was found that MEKC separated more fractions than CE, the MEKC method was optimised using voltage, SDS, β-CD and methanol concentrations, column temperature and buffer composition as variables (Figure 5.52). Borate buffer and alkaline pH produced more peaks than did phosphate buffer and acidic pH. It was found that CD-MEKC separated 53 peaks while HPLC separated only 40 UV-absorbing fractions. It was stated that the method is suitable for clinical diagnostic use and can be employed for the routine analysis of endogenous molecules in biological samples.

References

1. S. Terabe, K. Otsuka and H. Nishi, *J. Chromatogr. A*, 1994, **666**, 295–319.
2. M. Novotny, H. Soini and M. Stefansson, *Anal. Chem.*, 1994, **66**, 646A–655A.
3. M. M. Rogan, K. D. Altria and D. M. Goodall, *Chirality*, 1994, **6**, 25–40.
4. G. Gübitz and M. G. Schmid, *J. Chromatogr. A*, 1997, **792**, 179–225.
5. H. Nishi and S. Terabe, *J. Chromatogr. A*, 1995, **694**, 245–276.
6. T. J. Ward, *Anal. Chem.*, 1994, **66**, 633A–639A.

7. S. Fanali, *J. Chromatogr. A*, 1996, **735**, 77–121.

8. B. Chankvetadze, G. Endresz and G. Blaschke, *Chem. Soc. Rev.*, 1996, **25**, 141–153.

9. B. Chankvetadze, *J. Chromatogr. A*, 1997, **792**, 269–295.

10. S. Fanali, *J. Chromatogr. A*, 1997, **792**, 227–267.

11. I. S. Lurie, *J. Chromatogr. A*, 1997, **792**, 297–307.

12. J. H. T. Luong and A. L. Nguyen, *J. Chromatogr. A*, 1997, **792**, 431–444.

13. K. L. Larsen and W. Zimmermann, *J. Chromatogr. A*, 1999, **836**, 3–24.

14. Y.-Z. Hsieh and H.-Y. Huang, *J. Chromatogr. A*, 1996, **745**, 217–223.

15. M. Miura, Y. Terashita, K. Funazo and M. Tanaka, *J. Chromatogr. A*, 1999, **846**, 359–367.

16. O. Zerbinati, F. Trotta, C. Giovannoli, C. Baggiani, G. Giraudi and A. Vanni, *J. Chromatogr. A*, 1998, **810**, 193–200.

17. M. W. F. Nielen, *J. Chromatogr.*, 1993, **637**, 81–90.

18. A. W. Garrison, P. Schmitt and A. Kettrup, *J. Chromatogr. A*, 1994, **688**, 317–327.

19. Y. Mechref and Z. El Rassi, *Anal. Chem.*, 1996, **68**, 1771–1777.

20. S. K. Yeo, H. K. Lee and S. F. Y. Li, *J. Chromatogr.*, 1992, **594**, 335–340.

21. D. Kaniansky, E. Kremova, V. Madajova, M. Masar, J. Marak and F. I. Onuska, *J. Chromatogr. A*, 1997, **772**, 327–337.

22. C. A. Lucy, R. Brown and K. K.-C. Yeung, *J. Chromatogr. A*, 1996, **745**, 9–15.

23. A.-L. Nguyen and J. H. T. Luong, *Anal. Chem.*, 1997, **69**, 1726–1731.

24. K. Bächmann, A. Bazzanella, I. Hang and K.-Y. Han, *Fresenius J. Anal. Chem.*, 1997, **357**, 32–36.

25. O. H. J. Szolar, R. S. Brown and J. H. T. Luong, *Anal. Chem.*, 1995, **67**, 3004–3010.

26. R. S. Brown, J. H. T. Luong, O. H. J. Szolar, A. Halasz and J. Hawari, *Anal. Chem.*, 1996, **68**, 287–292.

27. Ph. Baumy, Ph. Morin, M. Dreux, M. C. Viaud, S. Boye and G. Guillaumet, *J. Chromatogr. A*, 1995, **707**, 311–326.

28. P. Gareil, J. P. Gramond and F. Guyon, *J. Chromatogr.*, 1993, **615**, 317–325.

29. S. Fanali and E. Camera, *Chromatographia*, 1996, **43**, 247–253.

30. G. G. Yowell, S. D. Fazio and R. V. Vivilecchia, *J. Chromatogr. A*, 1996, **745**, 73–79.

31. H. Wan and L. G. Blomberg, *J. Chromatogr. Sci.*, 1996, **34**, 540–546.

32. S. Cladrowa-Runge and A. Rizzi, *J. Chromatogr. A*, 1997, **759**, 157–165.

33. S. Cladrowa-Runge and A. Rizzi, *J. Chromatogr. A*, 1997, **759**, 167–175.

34. S. A. C. Wren, *J. Chromatogr. A*, 1997, **768**, 153–159.

35. Y. Kurosu, K. Murayama, K. N. Shindo, Y. Shisa, Y. Satou, M. Senda and N. Ishioka, *J. Chromatogr. A*, 1998, **802**, 129–134.

36. M. Jung and E. Francotte, *J. Chromatogr. A*, 1996, **755**, 81–88.

37. K. D. Altria, P. Parkin and M. G. Hindson, *J. Chromatogr. B*, 1996, **686**, 103–110.

38. S. Sabah and G. K. E. Scriba, *J. Chromatogr. A*, 1998, **822**, 137–145.

39. S. Sabah and G. K. E. Scriba, *J. Chromatogr. A*, 1999, **833**, 261–266.

40. K. Verleysen, S. Sabah, G. Scriba, A. Chen and P. Sandra, *J. Chromatogr. A*, 1998, **824**, 91–97.

41. G. M. Robinson, E. W. Taylor, M. R. Smith and C. E. Lunte, *J. Chromatogr. B*, 1998, **705**, 314–350.

42. E. Francotte, L. Brandel and M. Jung, *J. Chromatogr. A*, 1997, **792**, 379–384.

43. I. E. Valkó, H. Sirén and M.-L. Riekkola, *J. Chromatogr. A*, 1996, **737**, 263–272.

44. C. Desiderio and S. Fanali, *J. Chromatogr. A*, 1995, **716**, 183–196.

45. A. Werner, T. Nassauer, P. Kiechle and F. Erni, *J. Chromatogr. A*, 1994, **666**, 375–379.

46. B. A. Ingelse, F. M. Everaerts, C. Desiderio and S. Fanali, *J. Chromatogr. A*, 1995, **709**, 89–98.

47. M. Yoshinaga and M. Tanaka, *J. Chromatogr. A*, 1995, **710**, 331–337.

48. M. Yoshinaga and M. Tanaka, *Anal. Chim. Acta*, 1995, **316**, 121–127.

49. M. Yoshinaga and M. Tanaka, *J. Chromatogr. A*, 1994, **679**, 359–365.

50. A. Guttman, S. Brunet, C. Jurado and N. Cooke, *Chirality*, 1995, **7**, 409–414.

51. S. G. Penn, G. Liu, E. T. Bergström, D. M. Goodall and J. S. Loran, *J. Chromatogr. A*, 1994, **680**, 147–155.

52. M. E. Biggin, R. L. Williams and G. Vigh, *J. Chromatogr. A*, 1995, **692**, 319–325.

53. H. Wan, P. E. Andersson, A. Engström and L. G. Blomberg, *J. Chromatogr. A*, 1995, **704**, 179–193.

54. H. Wan, A. Engström and L. G. Blomberg, *J. Chromatogr. A*, 1996, **731**, 283–292.

55. M. Miura, K. Funazo and M. Tanaka, *Anal. Chim. Acta*, 1997, **357**, 177–185.

56. K.-H. Gahm and A. M. Stalcup, *Anal. Chem.*, 1995, **67**, 19–25.

57. W. Lindner, B. Böhs and V. Seidel, *J. Chromatogr. A*, 1995, **697**, 549–560.

58. V. Pacáková, J. Suchónková and K. Stulik, *J. Chromatogr. B*, 1996, **681**, 69–76.

59. C. E. Sänger-van de Griend, K. Grönongsson and T. Arvidsson, *J. Chromatogr. A*, 1997, **782**, 271–279.

60. H. Wan and L. G. Blomberg, *J. Chromatogr. A*, 1997, **792**, 393–400.

61. S. Li and D. K. Lloyd, *J. Chromatogr. A*, 1994, **666**, 321–335.

62. S. Mayer and V. Schurig, *J. Liq. Chromatogr.*, 1993, **16**, 915–931.

63. F. Leliévre, P. Gareil, Y. Bahaddi and H. Galons, *Anal. Chem.*, 1997, **69**, 393–401.

64. S. Fanali and Z. Aturki, *J. Chromatogr. A*, 1996, **722**, 312–317.

65. J. C. Reijenga, B. A. Ingelse and F. M. Everaerts, *J. Chromatogr. A*, 1997, **792**, 371–378.

66. M. Friedberg and Z. K. Shihabi, *J. Chromatogr. B*, 1997, **695**, 193–198.

67. S. Fanali, C. Desiderio and Z. Aturki, *J. Chromatogr. A*, 1997, **772**, 185–194.

68. Y. Y. Rawjee and Gy. Vigh, *Anal. Chem.*, 1994, **66**, 619–627.

69. J. L. Haynes, III, S. A. Shamsi, F. O'Keefe, R. Darcey and I. M. Warner, *J. Chromatogr. A*, 1998, **803**, 261–271.

70. S. Fanali and E. Camera, *Chromatographia*, 1996, **43**, 247–253.

71. Y. Y. Rawjee, D. U. Staerk and G. Vigh, *J. Chromatogr.*, 1993, **635**, 291–306.

72. Y. Y. Rawjee, R. L. Williams and G. Vigh, *J. Chromatogr. A*, 1994, **680**, 599–607.

73. Y. Y. Rawjee, R. L. Williams and G. Vigh, *Anal. Chem.*, 1994, **66**, 3777–3781.

74. J. B. Vincent and Gy. Vigh, *J. Chromatogr. A*, 1998, **816**, 233–241.

75. M. Fillet, L. Fotsing and J. Crommen, *J. Chromatogr. A*, 1998, **817**, 113–119.

76. F. Leliévre, C. Yan, R. N. Zare and P. Gareil, *J. Chromatogr. A*, 1996, **723**, 145–156.

77. F. Leliévre, P. Gareil and A. Jardy, *Anal. Chem.*, 1997, **69**, 385–392.

78. M. Fillet, I. Bechet, G. Schomburg, P. Hubert and J. Crommen, *J. High Resolut. Chromatogr.*, 1996, **19**, 669–673.

79. J. Szemán, N. Roos and K. Csabai, *J. Chromatogr. A*, 1997, **763**, 139–147.

80. G. Schulte, B. Chankvetadze and G. Blaschke, *J. Chromatogr. A*, 1997, **771**, 259–266.

81. M. Siluveru and J. T. Stewart, *J. Chromatogr. B*, 1997, **693**, 205–210.

82. B. Proksa, *J. Chromatogr. A*, 1998, **818**, 251–256.

83. T. Jira, A. Buntje and A. Karbaum, *J. Chromatogr. A*, 1998, **798**, 281–288.

84. A. Bunke and T. Jira, *J. Chromatogr. A*, 1998, **798**, 275–280.

85. A. Bunke and T. Jira, *Pharmazie*, 1996, **51**, 672–673.

86. Z. Wang, A. Huang, Y. Sun and Z. Sun, *J. High Resolut. Chromatogr.*, 1996, **19**, 697–699.

87. Z. Wang, A. Huang, Y. Sun and Z. Sun, *J. High Resolut. Chromatogr.*, 1996, **19**, 478–480.

88. S. Tahara, A. Okayama, Y. Kitada, T. Watanabe, H. Nakazawa, K. Kakehi and Y. Hisamatu, *J. Chromatogr. A*, 1999, **848**, 465–471.

89. B. Chankvetadze, G. Schulte, D. Bergenthal and G. Blaschke, *J. Chromatogr. A*, 1998, **798**, 315–323.

90. G. Schulte, S. Heitmeier, B. Chankvetadze and G. Blaschke, *J. Chromatogr. A*, 1998, **800**, 77–82.

91. B. Chankvetadze, G. Endresz, D. Bergenthal and G. Blaschke, *J. Chromatogr. A*, 1995, **717**, 245–253.

92. S. Paris, G. Blaschke, M. Locher, H. O. Borbe and J. Engel, *J. Chromatogr. B*, 1997, **691**, 463–471.

93. J. Liu, H. Coffey, D. J. Detlefsen, Y. Li and M. S. Lee, *J. Chromatogr. A*, 1997, **763**, 261–269.

94. P. K. Owens, A. F. Fell, M. W. Coleman and J. C. Berridge, *J. Chromatogr. A*, 1998, **797**, 149–164.

95. A. M. Abushoffa and B. J. Clark, *J. Chromatogr. A*, 1995, **700**, 51–58.

96. T. Horimai, M. Ohara and M. Ichinose, *J. Chromatogr. A*, 1997, **760**, 235–244.

97. C.-E. Lin, W.-C. Lin, W.-C. Chiou, E. C. Lin and C.-C. Chang, *J. Chromatogr. A*, 1996, **755**, 261–269.

98. S. Fanali and E. Camera, *J. Chromatogr. A*, 1996, **745**, 17–23.

99. P. Castelnovo and C. Albanesi, *J. Chromatogr. A*, 1996, **741**, 123–130.

100. C. Felli, G. Carrea, M. Chiari, M. De Amici and C. De Micheli, *J. Chromatogr. A*, 1996, **741**, 287–294.

101. K.-H. Gahm and A. M. Stalcup, *Chirality*, 1996, **8**, 316–324.

102. S. K. Branch, U. Holzgrabe, T. M. Jefferies, H. Mallwitz and F. J. R. Oxley, *J. Chromatogr. A*, 1997, **758**, 277–292.

103. N. Roos, K. Ganzler, J. Szemán and S. Fanali, *J. Chromatogr. A*, 1997, **782**, 257–269.

104. D. K. Bempong and I. H. Honigberg, *J. Pharm. Biomed. Anal.*, 1996, **15**, 233–239.

105. M. G. Schmidt, K. Wirnsberger and G. Gübitz, *Pharmazie*, 1996, **51**, 852–854.

106. H. Cai and G. Vigh, *J. Chromatogr. A*, 1998, **827**, 121–132.

107. M. Fillet, I. Bechet, P. Hubert and J. Crommen, *J. Pharm. Biomed. Anal.*, 1996, **14**, 1107–1114.

108. J. Sevcik, Z. Stránsky, B. A. Ingelse and K. Lemr, *J. Pharm. Biomed. Anal.*, 1996, **14**, 1089–1094.

109. H. L. Wu, K. Otsuka and S. Terabe, *J. Liq. Chromatogr. Relat. Technol.*, 1996, **19**, 1567–1577.

110. E. Szökö, J. Gyimesi, L. Barcza and K. Magyar, *J. Chromatogr. A*, 1996, **745**, 181–187.

111. A. Aumatell, R. J. Wells and D. K. Y. Wong, *J. Chromatogr. A*, 1994, **686**, 293–307.

112. I. S. Lurie, R. F. X. Klein, T. A. Dal Cason, M. J. LeBelle, R. Brenneisen and R. E. Weinberger, *Anal. Chem.*, 1994, **66**, 4019–4026.

113. E. Varesio and J.-L. Veuthey, *J. Chromatogr. A*, 1995, **717**, 219–228.

114. Z. Wang, Y. Sun and Z. Sun, *J. Chromatogr. A*, 1996, **735**, 295–301.

115. S. Cladrowa-Runge, R. Hirz, E. Kanndler and A. Rizzi, *J. Chromatogr. A*, 1995, **710**, 339–345.

116. G. Veseloh, H. Bartsch and W. A. König, *J. Microcolumn Sep.*, 1995, **7**, 355–363.

117. M. W. F. Nielen, *Anal. Chem.*, 1993, **65**, 885–893.

118. A. Aumatell and A. Guttman, *J. Chromatogr. A*, 1995, **717**, 229–234.

119. Z. Aturki and S. Fanali, *J. Chromatogr. A*, 1994, **680**, 137–146.

120. L. A. St. Pierre and K. B. Sentell, *J. Chromatogr. B*, 1994, **657**, 291–300.

121. H. Soini, M.-L. Riekkola and M. V. Novotny, *J. Chromatogr.*, 1992, **608**, 265–274.

122. B. Koppenhoefer, U. Epperlein, B. Christian, B. Lin, Y. Ji and Y. Chen, *J. Chromatogr. A*, 1996, **735**, 333–343.

123. T. E. Peterson, *J. Chromatogr.*, 1993, **630**, 353–361.

124. A.-E. F. Nassar, F. J. Guarco, J. D. Stuart and W. M. Reuter, *J. Chromatogr. Sci.*, 1998, **36**, 19–22.

125. R. J. Tait, D. O. Thompson, V. J. Stella and J. F. Stobaugh, *Anal. Chem.*, 1994, **66**, 4013–4018.

126. M. Heuermann and G. Blaschke, *J. Chromatogr.*, 1993, **648**, 267–274.

127. H. Nishi, Y. Kokusenya, T. Miyamoto and T. Sato, *J. Chromatogr. A*, 1994, **659**, 449–457.

128. R. Porra, M. G. Quaglia and S. Fanali, *J. Chromatogr.*, 1993, **635**, 291–306.

129. D. Belder and G. Schomburg, *J. Chromatogr. A*, 1994, **666**, 351–365.

130. M. H. Lamoree, A. F. H. Sprang, U. R. Tjaden and J. van der Greef, *J. Chromatogr. A*, 1996, **742**, 235–242.

131. Z. Wang, A. Huang, Y. Sun and Z. Sun, *J. Chromatogr. A*, 1996, **749**, 300–303.

132. A. Andrisano, R. Gotti, V. Cavrini, V. Tumiatti, G. Felix and I. W. Wainer, *J. Chromatogr. A*, 1998, **803**, 189–195.

133. M. Siluveru and J. T. Stewart, *J. Chromatogr. B*, 1997, **691**, 217–222.

134. B. J. Fish, *J. Chromatogr. A*, 1997, **768**, 81–87.

135. B. Koppenhoefer, U. Epperlein, R. Schlunk, X. Zhu and B. Lin, *J. Chromatogr. A*, 1998, **793**, 153–164.

136. Y. Y. Rawjee, R. L. Williams and G. Vigh, *J. Chromatogr. A*, 1993, **652**, 233–245.

137. L. Bingcheng, J. Yibing, C. Yuying, U. Epperlein and B. Koppenhoefer, *Chromatographia*, 1996, **42**, 106–110.

138. S. G. Penn, D. M. Goodall and J. S. Loran, *J. Chromatogr.*, 1993, **636**, 149–152.

139. S. G. Penn, E. Bergström, D. M. Goodall and J. S. Loran, *Anal. Chem.*, 1994, **66**, 2866–2873.

140. Y. Tanaka, M. Yanagawa and S. Terabe, *J. High Resolut. Chromatogr.*, 1996, **19**, 421–433.

141. B. Chankvetadze, N. Burjanadze, G. Pintore, D. Streickmann, D. Bergenthal and G. Blaschke, *Chirality*, 1999, **11**, 635–644.

142. V. Lemesle-Lamache, M. Taverna, D. Wouessidjewe, D. Duchene and D. Ferrier, *J. Chromatogr. A*, 1996, **735**, 321–331.

143. H. Nishi, K. Nakamura, H. Nakai and T. Sato, *J. Chromatogr. A*, 1994, **678**, 333–342.

144. F. Wang and M. G. Khaledi, *Anal. Chem.*, 1996, **68**, 3460–3467.

145. R. Kuhn, C. Steinmetz, T. Bereuterm, P. Haas and F. Erni, *J. Chromatogr. A*, 1994, **666**, 367–373.

146. I. Le Potier, S. L. Tamisier-Karolak, Ph. Morin, F. Megel and M. Taverna, *J. Chromatogr. A*, 1998, **829**, 341–349.

147. E. Billiot, J. Wiang and I. M. Warner, *J. Chromatogr. A*, 1997, **773**, 321–329.

148. M. Fillet, I. Becher, P. Chiap, Ph. Hubert and J. Crommen, *J. Chromatogr. A*, 1995, **717**, 203–209.

149. S. Pálmarsdóttir and L.-E. Edholm, *J. Chromatogr. A*, 1994, **666**, 337–350.

150. A. Guttman and N. Cooke, *J. Chromatogr. A*, 1994, **680**, 157–162.

151. T. Schmitt and H. Engelhardt, *Chromatographia*, 1993, **37**, 475–481.

152. Y. Y. Rawjee, R. L. Williams, L. A. Buckingham and Gy. Vigh, *J. Chromatogr. A*, 1994, **688**, 273–282.

153. C. Quang and M. G. Khaledi, *J. Chromatogr. A*, 1995, **692**, 253–265.

154. T. Schmitt and H. Engelhardt, *J. Chromatogr. A*, 1995, **697**, 561–570.

155. J. Luksa, Dj. Josic, B. Podobnik, B. Furlan and M. Kremser, *J. Chromatogr. B*, 1997, 367–375.

156. J. B. Vincent and G. Vigh, *J. Chromatogr. A*, 1998, **817**, 105–111.

157. H. Katayama, Y. Ishihama and N. Asakawa, *J. Chromatogr. A*, 1997, **764**, 151–156.

158. V. C. Anigbogu, C. L. Copper and M. J. Stepaniak, *J. Chromatogr. A*, 1995, **705**, 343–349.

159. B. Chankvetadze, G. Endresz and G. Blaschke, *J. Chromatogr. A*, 1995, **700**, 43–49.

160. G. Endresz, B. Chankvetadze, D. Bergenthal and G. Blaschke, *J. Chromatogr. A*, 1996, **732**, 133–142.

161. J. Gu and R. Fu, *J. Chromatogr. A*, 1994, **667**, 367–370.

162. B. Koppenhoefer, U. Epperlein, B. Christian, J. Yibing, C. Yuying and L. Bingcheng, *J. Chromatogr. A*, 1995, **717**, 181–190.

163. A. M. Stalcup and K. H. Gahm, *Anal. Chem.*, 1996, **68**, 1360–1368.

164. G. N. Okafo, K. K. Rana and P. Camilleri, *Chromatographia*, 1994, **39**, 627–630.

165. E. C. Rickard and R. J. Bopp, *J. Chromatogr. A*, 1994, **680**, 609–621.

166. P. Barták, J. Sevcik, T. Adam, D. Friedecky, K. Lemr and Z. Stránsky, *J. Chromatogr. A*, 1998, **818**, 231–238.

167. J. Liu and S. F. Y. Li, *J. Liq. Chromatogr. Relat. Technol.*, 1996, **19**, 1697–1713.

168. Ph. Morin, D. Daguet, J. P. Coic and M. Dreux, *J. Chromatogr. A*, 1999, **837**, 281–287.

169. K. Kawamura, *J. Chromatogr. A*, 1998, **802**, 167–177.

170. Tadey and W. C. Purdy, *J. Chromatogr. B*, 1994, **657**, 365–372.

171. R. Roldan-Assad and P. Gareil, *J. Chromatogr. A*, 1995, **708**, 339–350.

172. M. Stefansson and M. Novotny, *J. Am. Chem. Soc.*, 1993, **115**, 11573–11580.

173. K.-L. Kuo and Y.-Z. Hsieh, *J. Chromatogr. A*, 1997, **768**, 334–341.

174. M. Masár, D. Kaniansky and V. Madajová, *J. Chromatogr. A*, 1996, **724**, 327–336.

175. S. Razee, A. Tamura and T. Masujima, *J. Chromatogr. A*, 1995, **715**, 179–188.

176. K.-H. Gahm, L. W. Chang and D. W. Armstrong, *J. Chromatogr. A*, 1997, **759**, 149–155.

177. J. H. T. Luong and Y. Guo, *J. Chromatogr. A*, 1998, **811**, 225–232.

178. M. G. Schmid, K. Wirnsberger, T. Jira, A. Bunke and G. Gübitz, *Chirality*, 1997, **9**, 153–156.

179. T. Jira, A. Bunke, M. G. Schmid and G. Gübitz, *J. Chromatogr. A*, 1997, **761**, 269–275.

180. A. Fridström, L. Nyholm, Th. Netscher, N. Lundell and K. E. Markides, *Chromatographia*, 1997, **44**, 313–319.

181. W. X. Huang, S. D. Fazio and R. V. Vivilecchia, *J. Chromatogr. A*, 1997, **781**, 129–137.

182. Ph. Morin, D. Bellessort, M. Greux, Y. Troin and J. Gelas, *J. Chromatogr. A*, 1998, **796**, 375–383.

183. S. Hamai and H. Sakurai, *Anal. Chim. Acta*, 1999, **402**, 53–58.

184. J. Fischer, P. Jandera and V. Stanek, *J. Chromatogr. A*, 1997, **772**, 385–396.

185. P. Jandera, J. Fischer, V. Stanek, M. Kucerová and P. Zvonicek, *J. Chromatogr. A*, 1996, **738**, 201–213.

186. M. Miura, K. Funazo and M. Tanaka, *Anal. Chim. Acta*, 1997, **357**, 177–185.

187. S. Hamai and H. Sakurai, *J. Chromatogr. A*, 1998, **800**, 327–332.

188. B. A. Williams and Gy. Vigh, *Anal. Chem.*, 1997, **69**, 4445–4451.

189. A. Nardi, A. Eliseev, P. Bocek and S. Fanali, *J. Chromatogr.*, 1993, **638**, 247–253.

190. A. Bergholdt, J. Overgaard, A. Colding and R. B. Frederiksen, *J. Chromatogr.*, 1993, **644**, 412–415.

191. M. Y. Khaled and H. M. McNair, *J. High Resolut. Chromatogr.*, 1996, **19**, 143–150.

192. I. E. Valkó, H. A. H. Billiet, J. Frank and K. Ch. A. M. Luyben, *J. Chromatogr. A*, 1994, **678**, 139–144.

193. W. X. Huang, H. Xu, S. D. Fazio and R. Vivilecchia, *J. Chromatogr. B*, 1997, **695**, 157–162.

194. N. W. Smith, *J. Chromatogr. A*, 1993, **652**, 259–262.

195. M. Masár, R. Bodor and D. Kaniansky, *J. Chromatogr. A*, 1999, **834**, 179–188.

196. B. Chankvetadze, G. Endresz and G. Blascke, *J. Chromatogr. A*, 1995, **704**, 234–237.

197. B. Chankvetadze, G. Schulte and G. Blascke, *J. Chromatogr. A*, 1996, **732**, 183–187.

198. B. Chankvetadze, G. Endresz, G. Schulte, D. Bergenthal and G. Blascke, *J. Chromatogr. A*, 1996, **732**, 143–150.

199. J. B. Vincent, A. D. Sokolowski, T. V. Nguyen and Gy. Vigh, *Anal. Chem.*, 1997, **69**, 4226–4233.

200. J. B. Vincent, D. M. Kirby, T. V. Nguyen and Gy. Vigh, *Anal. Chem.*, 1997, **69**, 4419–4428.

201. H. Cai, T. V. Nguyen and Gy. Vigh, *Anal. Chem.*, 1998, **70**, 580–589.

202. G. M. Janini, H. J. Issaq and G. M. Muschik, *J. Chromatogr. A*, 1997, **792**, 125–141.

203. H. Nishi and S. Terabe, *J. Chromatogr. A*, 1996, **735**, 3–27.

204. H. Nishi, *J. Chromatogr. A*, 1996, **735**, 57–76.

205. Ph. Schmitt, A. W. Garrison, D. Freitag and A. Kettrup, *J. Chromatogr. A*, 1997, **792**, 419–429.

206. Y. He and H. K. Lee, *J. Chromatogr. A*, 1996, **749**, 227–236.

207. I. Benito, J. M. Saz, M. L. Marina, J. Jiménez-Barbero, M. J. González and J. C. Diez-Masa, *J. Chromatogr. A*, 1997, **778**, 77–85.

208. M. L. Marina, I. Benito, J. C. Diez-Masa and M. J. González, *J. Chromatogr. A*, 1996, **752**, 265–270.

209. M. L. Marina, I. Benito, J. C. Diez-Masa and M. J. González, *Chromato-graphia*, 1996, **42**, 269–272.

210. W.-C. Lin, F.-C. Chang and C.-H. Kuei, *J. Microcolumn Sep.*, 1999, **11**, 231–238.

211. U. Krismann and W. Kleiböhmer, *J. Chromatogr. A*, 1997, **774**, 193–201.

212. C. J. Smith, J. Grainger and G. Patterson, Jr., *J. Chromatogr. A*, 1998, **803**, 241–247.

213. B. Jiménez, D. G. Patterson, J. Grainger, Z. Liu, M. J. González and M. L. Marina, *J. Chromatogr. A*, 1997, **792**, 411–418.

214. C. L. Copper and M. J. Stepaniak, *Anal. Chem.*, 1994, **66**, 147–154.

215. W. C. Brumley and W. J. Jones, *J. Chromatogr. A*, 1994, **680**, 163–173.

216. M. J. Stepaniak, C. L. Copper, K. W. Whitaker and V. C. Anigbogu, *Anal. Chem.*, 1995, **67**, 2037–2041.

217. R. Szücs, J. V. Beaman and A. M. Lipczynski, *J. Chromatogr. A*, 1999, **836**, 53–58.

218. S. Terabe, Y. Miyashita, Y. Ishihama and O. Shibata, *J. Chromatogr.*, 1993, **636**, 47–55.

219. Y. Liu, J. Gu and R. Fu, *J. High Resolut. Chromatogr.*, 1997, **20**, 159–164.

220. R. Furuta and T. Doi, *J. Chromatogr. A*, 1994, **676**, 431–436.

221. K. Otsuka and S. Terabe, *J. Liq. Chromatogr.*, 1993, **16**, 945–953.

222. A. Aumatell and R. J. Wells, *J. Chromatogr. A*, 1994, **688**, 329–337.

223. H.-Y. Huand, K.-L. Kuo and Y.-Z. Hsieh, *J. Chromatogr. A*, 1997, **771**, 267–274.

224. J. C. Archambault, P. Morin, P. André and M. Dreux, *Spectra Analyse*, 1997, **199**, 31–32.

225. U. B. Nair, K. L. Rundlett and D. W. Armstrong, *J. Liq. Chromatogr. Relat. Technol.*, 1997, **20**, 203–216.

226. H. Wan and L. G. Blomberg, *J. Chromatogr. A*, 1997, **758**, 303–311.

227. B. J. Spencer and W. C. Purdy, *J. Chromatogr. A*, 1997, **782**, 227–235.

228. L. N. Alfazema, M. E. P. Hows, S. Howells and D. Perrett, *Electrophoresis*, 1997, **18**, 1847–1856.

Subject Index